S0-CAT-745

Electronic Noise and Low Noise Design

Macmillan New Electronics Series
Series Editor: Paul A. Lynn

Rodney F. W. Coates, *Underwater Acoustic Systems*
W. Forsythe and R. M. Goodall, *Digital Control*
C. G. Guy, *Data Communications for Engineers*
Paul A. Lynn, *Digital Signals, Processors and Noise*
Paul A. Lynn, *Radar Systems*
A. F. Murray and H. M. Reekie, *Integrated Circuit Design*
F. J. Owens, *Signal Processing of Speech*
Dennis N. Pim, *Television and Teletext*
M. J. N. Sibley, *Optical Communications*
Martin S. Smith, *Introduction to Antennas*
P. M. Taylor, *Robotic Control*
G. S. Virk, *Digital Computer Control Systems*
Allan Waters, *Active Filter Design*

Series Standing Order

If you would like to receive future titles in this series as they are
published, you can make use of our standing order
facility. To place a standing order please contact your
bookseller or, in case of difficulty, write to us at the address
below with your name and address and the name of the
series. Please state with which title you wish to begin your
standing order. (If you live outside the UK we may not have
the rights for your area, in which case we will forward your
order to the publisher concerned.)

Customer Services Department, Macmillan Distribution Ltd,
Houndmills, Basingstoke, Hampshire, RG21 2XS, England

Electronic Noise
and
Low Noise Design

Peter J. Fish

MSc, FInstP, FIPSM, CEng, CPhys, Sen. Mem. IEEE
Senior Research Fellow, School of Electronic Engineering Science
University of Wales, Bangor
and
Consultant Medical Physicist
Gwynedd and Clwydd Health Authorities

150th YEAR

MACMILLAN

© Peter J. Fish 1993

All rights reserved. No reproduction, copy or transmission of
this publication may be made without written permission.

No paragraph of this publication may be reproduced, copied or
transmitted save with written permission or in accordance with
the provisions of the Copyright, Designs and Patents Act 1988,
or under the terms of any licence permitting limited copying
issued by the Copyright Licensing Agency, 90 Tottenham Court
Road, London W1P 9HE.

Any person who does any unauthorised act in relation to this
publication may be liable to criminal prosecution and civil
claims for damages.

First published 1993 by
THE MACMILLAN PRESS LTD
Houndmills, Basingstoke, Hampshire RG21 2XS
and London
Companies and representatives
throughout the world

ISBN 0-333-57309-9 hardcover
ISBN 0-333-57310-2 paperback

A catalogue record for this book is available
from the British Library.

Printed and bound in Great Britain by
Mackays of Chatham PLC, Chatham, Kent

To Kathryn and Marianne

Contents

Series Editor's Foreword

The rapid development of electronics and its engineering applications ensures that new topics are always competing for a place in university and polytechnic courses. But it is often difficult to find suitable books for recommendation to students, particularly when a topic is covered by a short lecture module, or as an 'option'.

This Series offers introductions to advanced topics. The level is generally that of second and subsequent years of undergraduate courses in electronic and electrical engineering, computer science and physics. Some of the authors will paint with a broad brush; others will concentrate on a narrower topic, and cover it in greater detail. But in all cases the titles in the Series will provide a sound basis for further reading of the specialist literature, and an up-to-date appreciation of practical applications and likely trends.

The level, scope and approach of the Series should also appeal to practising engineers and scientists encountering an area of electronics for the first time, or needing a rapid and authoritative update.

Paul A. Lynn

Preface

Noise is a problem in many electronic circuits and systems. Arising from the random movement of electrons in circuit elements (intrinsic noise) or from spuriously coupled signals from other circuits and systems (interference), it corrupts the signal of interest and introduces an uncertainty into the information it contains.

Intrinsic noise and interference are usually treated separately. The latter is normally the subject of books on electromagnetic compatibility (EMC). However the problems caused by both types of noise are similar and there is good reason for treating them together. Indeed it is often important for the design engineer to keep both types of noise in mind even when concentrating on one. For example, there is usually little point in incorporating shielding in a design in order to reduce interference well below the noise level determined by intrinsic noise.

This book covers both types of noise, and, within the category of interference, in addition to noise introduced by electric and magnetic fields, noise arising from the transduction of mechanical and thermal disturbances is described. In all cases the means of reducing noise to acceptable or minimum achievable levels are described.

The book aims to provide an introduction to the problem of noise from the viewpoint of a circuit designer, covering the theory of intrinsic noise, electromagnetic compatibility and the basis of low noise design. It will be of value to final year and postgraduate electronic engineering students taking courses on electronic noise or EMC, to postgraduate research students whose projects include low noise design and to practising engineers whose qualifying courses covered the subject inadequately or who need to refresh or improve their knowledge of this area of electronic engineering.

The author's interest in this subject arises from a 24-year involvement in medical instrumentation, dealing with low level signals in circuits with a wide range of impedance levels and from sub-audio to radio frequencies. The book, it is hoped, reflects this range, and all readers, whether concerned with signals from chemical sensors with time constants measured in seconds or with telemetry signals with bandwidths measured in megahertz, should find something of value.

A major group of noise signals is random in nature and, since these signals are often poorly understood, a chapter on the properties and the characterisation of random signals is included – providing theoretical background to the following chapters and to further reading. The various types of interference, both of electrical and non-electrical origin, and which are influenced strongly by the physical design of equipment, are described in chapter 3. Intrinsic noise, determined by circuit design, is covered in chapter 4. The methods of noise circuit analysis and noise models of common circuit elements are described in chapters 5 and 6 respectively. Chapter 7 covers the techniques of noise measurement and chapter 8 the use of industry-standard circuit-simulation software SPICE in intrinsic noise analysis. The lessons of the previous chapters, particularly 3 to 6, are brought together in chapter 9 which describes the methods of low noise design – covering the basic theory and techniques of electromagnetic compatibility and the methods of minimising intrinsic noise.

SI units are used throughout. In particular it should be noted that distances are in metres unless otherwise stated. The term *power-line* rather than *mains* is used and power-line frequencies of 50 and 60 Hz are used in examples and exercises. Negative exponents rather than '/' are used in most units. For example, intrinsic noise levels are expressed in $VHz^{-1/2}$ rather than $V/\sqrt{Hz}$.

Some of the exercises at the end of the book are adapted from my final year BEng Electronic Engineering exam questions. The permission of the University of Wales to use these questions is gratefully acknowledged.

My thanks are due also to many friends and colleagues who, often without realising it, imparted their enthusiasm for this Cinderella area of electronic engineering or gave me the opportunity to study it. I am particularly indebted to Dr Keith Battye who introduced me to the problem and challenge of noise in electromyography and who enabled me to study it in other areas of medical instrumentation, and to Professor John O'Reilly and Dr Peter Smith for giving me the opportunity to extend my interest, in teaching and production of this book. I should also like to thank Katie Parry for cheerfully typing much of the manuscript, Tony Griffiths for his artistic expertise and my final year students who, without knowing it, encouraged my production of this book in the hope that it might improve their exam grades! Special thanks are due to my family who tolerated my absence from normal family activities over a considerable period of time.

1 Introduction

1.1 Definition

There are at least two commonly used definitions of electronic noise. Noise is either a random fluctuation in potential difference or current resulting from the random movement of charge carriers, or it is an unwanted signal tending to obscure or interfere with a required signal. This second definition is broader and includes the random noise but it also includes interference – that is, the addition to the 'required' signal of other signals which may or may not be random. An example of a non-random or deterministic noise signal is mains (line) hum in an audio system. This is a 50 or 60 Hertz signal (together with harmonics) introduced into the audio signal path by means of magnetic induction from a transformer or capacitative coupling between the power supply and audio signal leads. Another example is the reception of broadcast signals by part of a circuit acting as an aerial. In both these cases the interfering signal is a 'wanted' signal in some circuits but 'unwanted' and interfering if added to another signal. In this text we shall adopt the broad definition.

1.2 Noise categories

Electronic noise may be conveniently divided into two categories determined by the noise source. In the first category the noise source is external to the elements of the circuit under consideration. In the second category the noise is generated within the circuit elements and arises from the random motion of charge carriers. This second category of noise is sometimes called 'intrinsic' noise.

Noise in the first category (external noise source) may be random but is more usually of a deterministic nature. The noise is generated in some external source and is coupled into the circuit of interest. This noise, for example, can be: electromagnetic radiation from transmitters or from arcing in brush motors and switch contacts picked up by part of the circuit of interest acting as an aerial; the alternating magnetic field from a transformer or motor inducing a current of a similar frequency within conductors in the circuit; noise signals capacitatively coupled to nearby conductors; and even mechanical vibration coupled into the circuit by means of cable movement. Noise in this category is strongly influenced by the layout and construction of the apparatus concerned. The length and positioning of conductors are critical and the noise may be minimised by attention to physical design.

Most of the problems in this category are concerned with noise coupled by magnetic and electric fields (including conducted noise). The field of study of this type of noise and the techniques of reduction is called *electromagnetic compatibility* (EMC).

Noise in the second category – intrinsic noise – is caused by the random movement of charge carriers within conductors. Shot noise is the random component of current flowing across a potential barrier – in a semiconductor junction for example – and results from the random crossing of the potential barrier by the charge carriers. Thermal noise, which is present even in the absence of current flow, results from the thermally induced random motion of charge carriers within a conductor and is dependent on temperature. The level of this category of noise is governed by the values of the circuit components – for example, resistor values, their material composition and the current passing through them – and is therefore governed by circuit design rather than circuit layout.

1.3 Effect of noise

The effect of noise clearly depends on the function of the circuit being affected. Noise added to a signal may give rise to an error in the measurement of the signal level. For example, consider a noise-contaminated signal from a sensor monitoring changes in the sensed quantity (temperature, chemical concentration, etc.) as shown in figure 1.1. The sampled signal indicates erroneous levels as shown.

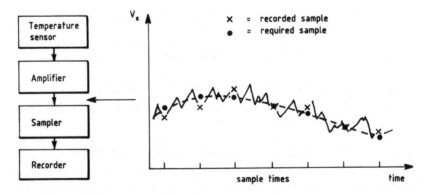

Figure 1.1   Temperature recording: – – – noise-free temperature
signal, ——— noisy temperature signal

Noise may give rise to errors in the measurement of signal timing or phase. For example, the measurement of distance using pulse-echo techniques, used in radar, sonar, ultrasonic flaw detection and medical ultrasound imaging, will be affected. The echo signal from an ultrasonic pulse-echo instrument is shown in figure 1.2. The distance of the reflector from the transducer may be measured from the time of arrival of the leading edge of the received pulse – specifically when it exceeds a pre-set threshold. The presence of noise may cause premature or late threshold crossing leading to an erroneous measurement of pulse-echo delay and therefore reflector depth.

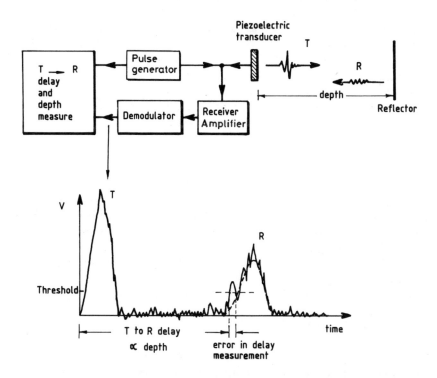

Figure 1.2 Ultrasonic depth measurement. T – transmitter pulse, R – received pulse. Dashed line – noise-free received pulse

Data in a digital format may be corrupted by noise. For example, the synchronous receiver of a binary signal will assign '0' or '1' levels to the sampled signal depending on the signal level with respect to a fixed threshold (figure 1.3). The presence of noise may lead to erroneous assignment of '0' or '1' and, therefore, corruption of the coded information.

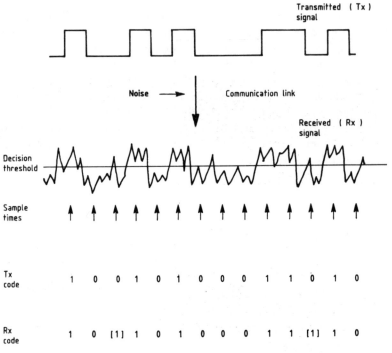

Figure 1.3   Noise-induced error in binary code

Noise in conjunction with an audio signal may diminish pleasure when the audio signal is a piece of music or lead to misunderstanding if the signal contains a message.

A noisy video signal will lead to masking of detail on the video image (figure 1.4) and, if the signal/noise ratio is very poor, give rise to loss of synchronisation of the video time-bases – leading to 'tearing' of the picture.

1.4   Low noise design

The low priority given to noise in many electronics courses can lead to a design philosophy which leaves noise problems to be 'sorted out' later when all the other circuit design problems have been solved. The designer may be lucky, and the circuit may require merely the repositioning of a few leads and the addition of a capacitor to get it working. The difficulty with this approach is that it can easily lead to a circuit which is unusable as a result of noise and which requires major, not minor, alterations to make it work.

A far better approach, which is adopted by successful designers in the electronics industry, is to consider noise performance right from the start. Seeing it as an important design constraint can lead to the avoidance of many problems.

Image   Video line

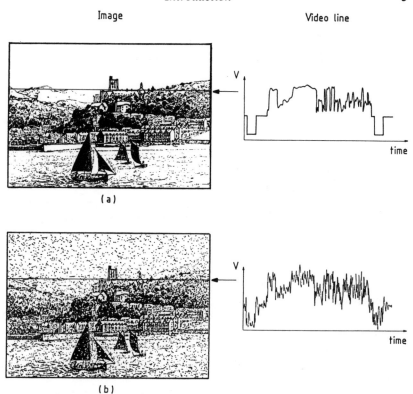

Figure 1.4   Video image and signal: (a) noise-free; (b) with added noise.
(Reproduced by permission of the University of Wales,
Bangor)

Since electrical noise is usually at a low level, it is normally a problem in
circuits where the signal level is also low. It is tempting under these
conditions simply to amplify the signal – but, of course, this merely
increases both signal and noise and does not improve the situation. It is
important to think not of design for high signal level or for low noise level
but in terms of a good signal/noise ratio.

The aim of this text is to provide an introduction to the problem of noise
as it relates to the design of electronic circuits. The physical basis for the
noise will not be described in detail except where it is important in a design
context. The reader will be guided to texts covering more fundamental
aspects of noise where necessary. In addition, the more advanced signal
processing techniques for reducing the effects of noise will not be discussed.
These points are adequately covered in many books on signal processing.
This book is concerned with the basics of designing with the problem of
noise in mind.

2 Random Signals

2.1 Introduction

A random, or stochastic signal is characterised by its lack of predictability. Unlike a deterministic signal whose value at any instant is known, the value of a random signal at any instant is not known and cannot be predicted from a knowledge of values at other past instants. The characteristics of random signals can be described only in terms of averages. In this chapter some of the properties and methods of characterisation of random signals are described. The treatment is in no sense comprehensive. Only those results that are required for an understanding of the aspects of noise analysis described in this book are covered here. Other books giving a more exhaustive treatment are listed in the Bibliography. Note that some equations quoted are valid only for the *real* time-functions used in the rest of the book.

2.2 Elements of probability

2.2.1 Probability of events

The outcome of any experiment is governed in some degree by chance. An extreme example is the number displayed when a die is thrown or the face displayed when a coin is tossed. We cannot predict the outcome at the start of the experiment – it is determined by chance. It is possible, however, to predict, to some degree of accuracy, the relative frequency of the different results from a large number of experiments. As a result of a large number of coin tosses (experiments) approximately half the results will be heads and half tails and it is found that this approximation becomes more accurate as the number of experiments increases.

In general, if the total number of experiments is N and the number of results of type A is m_A then the relative frequency of result A is m_A/N and the probability of result A is:

$$P(A) = \lim_{N \to \infty} \frac{m_A}{N} \tag{2.1}$$

If the outcome of the experiment can be $A_1, A_2, A_3 \ldots$ or A_N then provided each outcome is mutually exclusive – that is, any one result precludes the possibility of any other – then the probability of either A_1 or A_2 occurring is:

$$P(A_1 + A_2) = \lim_{N \to \infty} \frac{m_{A_1} + m_{A_2}}{N}$$

$$= P(A_1) + P(A_2) \tag{2.2}$$

and the probability of all results occurring is:

$$P(A_1 + A_2 + \ldots A_M) = \sum_{i=1}^{M} P(A_i)$$

$$= \lim_{N \to \infty} \frac{\sum\limits_{i=1}^{M} m_{A_i}}{N}$$

$$= 1 \tag{2.3}$$

denoting a certainty.

2.2.2 *Conditional probability and independence*

In some experiments the probability of one outcome is influenced by the outcome of another experiment. For example, the probability of drawing an ace from a pack of cards is reduced if an ace is drawn during a previous experiment and not replaced. The probability of the outcome of one experiment is conditional on the result of a previous experiment.

If outcome A occurs n_A times out of N trials and out of those n_A there is a joint occurrence of outcomes A and B n_{AB} times, then the probability of joint occurrence of A and B is:

$$P(AB) = \lim_{N \to \infty} \frac{n_{AB}}{N}$$

$$= \lim_{N \to \infty} \frac{n_A}{N} \frac{n_{AB}}{n_A}$$

But:

$$\lim_{N \to \infty} \frac{n_A}{N} = P(A)$$

and:

$$\lim_{N \to \infty} \frac{n_{AB}}{n_A} = P(B \mid A)$$

where $P(B \mid A)$ is the probability of outcome B given the outcome A or the conditional probability of B given A.

Therefore we may write:

$$P(AB) = P(A)P(B \mid A)$$

or:

$$P(B \mid A) = \frac{P(AB)}{P(A)} \qquad (2.4)$$

similarly:

$$P(A \mid B) = \frac{P(AB)}{P(B)} \qquad (2.5)$$

Eliminating $P(AB)$ we obtain:

$$P(A \mid B)P(B) = P(B \mid A)P(A) \qquad (2.6)$$

which is known as Bayes' theorem.

Note that if A and B are independent – that is, one outcome has no effect on the other then:

$$P(AB) = P(A)P(B) \qquad (2.7)$$

2.2.3 Discrete random variable

If the various outcomes of an experiment have associated numerical values x_i $(i = 1, 2 \ldots M_x)$ and the value x_i occurs n_i times in a total number of values N, then the probability of x_i occurring is:

$$P(x_i) = \lim_{N \to \infty} \frac{n_i}{N} \qquad (2.8)$$

and may be shown graphically as in figure 2.1.

The quantity x is known as a discrete random variable. The equation corresponding to equation (2.2) is that giving the probability of x taking any values between and including x_{m_1} and x_{m_2}, that is:

$$P(x_{m_1} \leqslant x \leqslant x_{m_2}) = \sum_{i=m_1}^{m_2} P(x_i) \qquad (2.9)$$

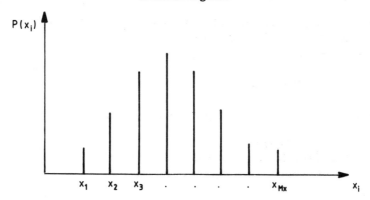

Figure 2.1 Probability function of a discrete random variable

and of course:

$$\sum_{i=1}^{M_x} P(x_i) = 1 \tag{2.10}$$

The joint probability of two discrete random variables x_i and y_j is written $P(x_i, y_j)$ – the probability of x_i and y_j occurring simultaneously. If x_i and y_j are independent then:

$$P(x_i, y_j) = P(x_i)P(y_j) \tag{2.11}$$

The equation corresponding to (2.9) for the two variables, that is the probability that x and y are simultaneously in the ranges $x_{m_1} \rightarrow x_{m_2}$ and $y_{n_1} \rightarrow y_{n_2}$ respectively is:

$$P(x_{m_1} \leqslant x \leqslant x_{m_2}, y_{n_1} \leqslant y \leqslant y_{n_2}) = \sum_{i=m_1}^{m_2} \sum_{j=n_1}^{n_2} P(x_i, y_j) \tag{2.12}$$

and the equation corresponding to (2.10) is:

$$\sum_{i=1}^{M_x} \sum_{j=1}^{N_y} P(x_i, y_j) = 1 \tag{2.13}$$

2.2.4 Continuous random variable

In many cases, x is a continuous random variable. It may be, for example, a measurement in which there is a degree of uncertainty such that a number of measurements lead to a range of values of x. The range of possible values of x may be subdivided into a number of classes of width Δx. If m_n is the number of measurements in the range $x_n - \Delta x/2$ to $x_n + \Delta x/2$ out of a total number of measurements N, then the probability of finding a measurement in this range is:

$$P(x_n - \Delta x/2 < x \leqslant x_n + \Delta x/2) = P_\Delta(x_n) = \lim_{N\to\infty} \frac{m_n}{N} \qquad (2.14)$$

This probability is shown, as a histogram, in figure 2.2. The probability of a measurement falling in the range of classes $n_1 \to n_2$ is:

$$\lim_{N\to\infty} \frac{1}{N} \sum_{n=n_1}^{n_2} m_n = \sum_{n=n_1}^{n_2} P_\Delta(x_n) \qquad (2.15)$$

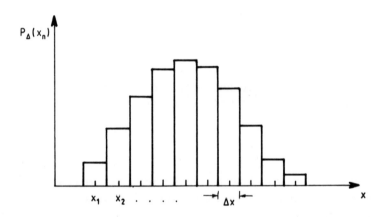

Figure 2.2   Probability of continuous random variable (x) falling in range Δx centred on x_n

This method of expressing the variation of probability with x is limited because of its dependence on the class width Δx. As Δx is reduced, for example (number of classes increases), the probability of a measurement being in any one class reduces. A more useful way of expressing the probability variation with x is the probability density function:

$$p(x) = \lim_{\Delta x \to 0} \frac{P_\Delta(x_n)}{\Delta x} \qquad (2.16)$$

This is shown in figure 2.3. The probability of a measurement falling in the range $x_1 \to x_2$ is:

$$P(x_1 < x \leqslant x_2) = \int_{x_1}^{x_2} p(x)\, dx \tag{2.17}$$

and is given by the hatched area in figure 2.3. The area under the whole curve represents the total probability or unity. Thus:

$$\int_{-\infty}^{\infty} p(x)\, dx = 1 \tag{2.18}$$

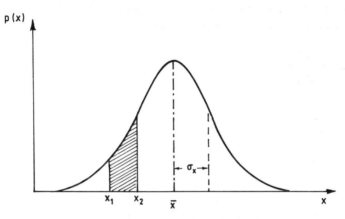

Figure 2.3   Probability density function of continuous random variable x

The corresponding equations for continuous random variables with a joint probability density function $p(x, y)$ are:

$$P(x_1 < x \leqslant x_2, y_1 < y \leqslant y_2) = \int_{x_1}^{x_2} \int_{y_1}^{y_2} p(x, y)\, dx\, dy \tag{2.19}$$

$$\int_{-\infty}^{\infty} \int_{-\infty}^{\infty} p(x, y)\, dx\, dy = 1 \tag{2.20}$$

As in the case of discrete variables, if x and y are independent then:

$$p(x, y) = p(x)p(y) \tag{2.21}$$

2.2.5 Means

The mean or average value of a set of values of a discrete random variable is defined as the sum of the values divided by the number in the set. We get an increasingly consistent figure for the mean by letting the set number increase. Thus the average is:

$$\bar{x} = \lim_{N\to\infty} \frac{1}{N} \sum_{i=1}^{M_x} x_i n_i$$

$$= \sum_{i=1}^{M_x} x_i P(x_i) \tag{2.22}$$

For a continuous variable this becomes:

$$\bar{x} = \int_{-\infty}^{\infty} x p(x) dx \tag{2.23}$$

Another useful characteristic figure of a random variable is its root mean square deviation from the mean – known as the standard deviation. This gives a measure of the range of values of x. Following a similar argument to that above, the mean square deviation or variance is:

$$\sigma_x^2 = \overline{(x - \bar{x})^2} = \int_{-\infty}^{\infty} (x - \bar{x})^2 p(x)\, dx \tag{2.24}$$

and the rms, or standard, deviation:

$$\sigma_x = \left[\int_{-\infty}^{\infty} (x - \bar{x})^2 p(x)\, dx \right]^{1/2} \tag{2.25}$$

and for the special case of a zero-mean variable:

$$\sigma_x = \left(\overline{(x - \bar{x})^2} \right)^{1/2} = \left(\overline{x^2} \right)^{1/2} = \left[\int_{-\infty}^{\infty} x^2 p(x)\, dx \right]^{1/2} \tag{2.26}$$

This is the root mean square (rms) value of x.

For jointly random variables x, y, the mean or expected value of their product is:

$$\overline{xy} = \int_{-\infty}^{\infty} \int_{-\infty}^{\infty} xy\, p(x,y)\, dx\, dy \tag{2.27}$$

If x and y are independent then:

$$p(x,y) = p(x)p(y) \tag{2.28}$$

and (2.27) becomes:

$$\overline{xy} = \int_{-\infty}^{\infty} x p(x)\, dx \int_{-\infty}^{\infty} y p(y)\, dy = \bar{x}\,\bar{y} \tag{2.29}$$

2.2.6 Gaussian probability density function

Many random variables have a Gaussian probability density function, given by:

$$p(x) = \frac{1}{\sigma_x (2\pi)^{1/2}} \exp\left(0.5(x - \bar{x})^2 / \sigma_x^2\right) \tag{2.30}$$

It is easily verified that this distribution obeys equations (2.18), (2.23) and (2.24).

An example of this form of probability density function (pdf) is shown in figure 2.3.

2.2.7 Central limit theorem

Many random variables are the large-scale manifestation of large numbers of small-scale random events. It has been shown that the pdf of the sum of a number of random variables tends to a Gaussian as the number of random variables increases and this is independent of the pdfs of the constituent variables. This is known as the *central limit theorem*.

2.2.8 Correlation

Consider the sum z of random variables x and y.

$$z = x + y \tag{2.31}$$

The variance of z is:

$$\sigma_z^2 = \overline{(z - \bar{z})^2} = \overline{(x - \bar{x})^2} + \overline{(y - \bar{y})^2} + \overline{2(x - \bar{x})(y - \bar{y})}$$
$$= \sigma_x^2 + \sigma_y^2 + \overline{2(x - \bar{x})(y - \bar{y})} \tag{2.32}$$

or:

$$\bar{z}^2 = \bar{x}^2 + \bar{y}^2 + 2\overline{xy}$$

for zero-mean variables.

The value of $\overline{(x - \bar{x})(y - \bar{y})}$ clearly depends on the relationship between x and y. If x and y are independent, then for each value of $(x - \bar{x})(y - \bar{y})$ there is a value $-(x - \bar{x})(y - \bar{y})$ of equal probability and $\overline{(x - \bar{x})(y - \bar{y})}$ equals zero. On the other hand, if x and y are linearly dependent, that is:

$$(y - \bar{y}) = k(x - \bar{x}) \tag{2.33}$$

then:

$$\overline{(x-\bar{x})(y-\bar{y})} = k\overline{(x-\bar{x})^2} = k\sigma_x^2$$

or:

$$\overline{(x-\bar{x})(y-\bar{y})} = \overline{(y-\bar{y})^2}/k = \sigma_y^2/k \tag{2.34}$$

or, combining the above:

$$\overline{(x-\bar{x})(y-\bar{y})} = \text{sign}(k)\sigma_x\sigma_y \tag{2.35}$$

We define a correlation coefficient:

$$\rho = \frac{\overline{(x-\bar{x})(y-\bar{y})}}{\sigma_x\sigma_y} \tag{2.36}$$

which ranges from -1 to $+1$. In the case of a linear relationship between x and y, ρ is ± 1 and when they are independent ρ is zero. The value of ρ defines the degree of correlation between the variables. Equation (2.32) now becomes:

$$\sigma_z^2 = \sigma_x^2 + \sigma_y^2 + 2\rho\sigma_x\sigma_y \tag{2.37}$$

The relationships between x and y for various values of ρ are shown in figure 2.4.

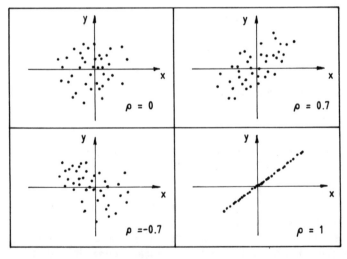

Figure 2.4   Plot of corresponding samples of two random variables for different degrees of correlation

If y is the sum of two variables, one independent of x and one proportional to x, then the value of ρ depends on the relative proportions of the two components. Note that the condition of independence of x and y for $\rho = 0$ is sufficient but not necessary. It is possible to have random variables which are uncorrelated but dependent – for example

$$x = \sin(\theta)$$
$$y = \cos(\theta) \tag{2.38}$$

where θ is a random variable.

2.2.9 Measurement errors

The random variable of interest is often a measurement or estimate of a quantity. If the measurement is unbiased – that is, the average measurement is equal to the true value of the quantity being measured – then the measurement error is usually specified by the rms deviation of the measurements from the mean, or standard deviation σ. The error is often expressed as a fraction of the mean. For example:

$$\epsilon_x = \sigma_x / \bar{x}$$

If a quantity z is calculated from the sum of two measured quantities x and y with independent measurement errors, then:

$$\bar{z} = \bar{x} + \bar{y}$$

and, from (2.37), the fractional error in z is:

$$\epsilon_z = \sigma_z / \bar{z} = \frac{(\sigma_x^2 + \sigma_y^2)^{1/2}}{\bar{z}} \tag{2.39}$$

Note that if the errors in x and y are equal then:

$$\epsilon_z = 2^{1/2} \sigma_x / \bar{z} \tag{2.40}$$

Note also that if one of the errors is only a small factor times the other then the smaller error can often be ignored. For example, if:

$$\sigma_y = 3\sigma_x \tag{2.41}$$

then:

$$\sigma_z = 1.054\sigma_y \tag{2.42}$$

and only a 5.4 per cent error results from ignoring σ_x.

If either $\bar{x}$ or $\bar{y}$ is negative and $|\bar{x}|$ and $|\bar{y}|$ are close, then the fractional error in z can be high. For example, if the errors in x and y are equal and:

$$\bar{y} = -0.8\bar{x}$$

then:

$$\epsilon_z = \frac{2^{1/2}\sigma_x}{0.2\bar{x}} \simeq 7\epsilon_x \qquad (2.43)$$

2.3 Random processes

2.3.1 Introduction

If the outcome of an experiment is not just a single number but a random function of time, then this outcome is called a random process and the set of random time-functions resulting from a number of experiments is known as an ensemble. A sample ensemble of five random time-functions from a random process is shown in figure 2.5. For example, we shall see later that

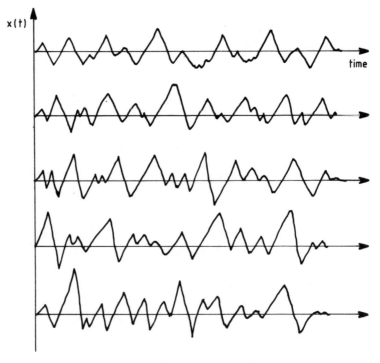

Figure 2.5   An ensemble of five sample time-functions of a random process

the statistical properties of thermal noise from a resistor is defined by its resistance and by the frequency band in which the noise is measured. The noise voltage waveforms resulting from bandpass filtering the thermal noise from a number of similar resistors would constitute an ensemble of time-functions. Often, however, there is only one experiment and one time-function and the ensemble remains conceptual. Note that the ensemble is a sample from the infinity of time-functions which constitute the random process.

2.3.2 Means

At any time t, the ensemble values of the random functions constitute a random variable and the discussion in the previous sections apply to this variable. For example, if the probability density function at time t is $p(x; t)$ then the mean or expected values of x and x^2 are:

$$\bar{x}(t) = \mathcal{E}[x(t)] = \int_{-\infty}^{\infty} x p(x; t) \, dx \qquad (2.44)$$

and:

$$\overline{x^2}(t) = \mathcal{E}[x^2(t)] = \int_{-\infty}^{\infty} x^2 p(x; t) \, dx \qquad (2.45)$$

(where $\mathcal{E}$ is the expected value operator) and, in general, are functions of time.

2.3.3 Autocorrelation and cross-correlation functions

It is often useful to know how rapidly the time-functions of a random process change with time and a convenient measure of this is the autocorrelation function:

$$R_x(t_1, t_2) = \mathcal{E}[x(t_1)x(t_2)] = \overline{x(t_1)x(t_2)}$$

$$= \int \int x_1 x_2 p(x_1, x_2; t_1, t_2) \, dt_1 dt_2 \qquad (2.46)$$

When t_1 equals t_2 this becomes $\overline{x^2(t_1)}$ – the mean square value of x at time t_1. As the interval $t_2 - t_1$ increases then the degree of correlation between $x(t_1)$ and $x(t_2)$ decreases and in the limit:

$$\lim_{t_2 - t_1 \to \infty} R_x(t_1, t_2) = \overline{x(t_1)} \; \overline{x(t_2)} \qquad (2.47)$$

which equals zero for a zero-mean process.

The width of the autocorrelation function clearly depends on the rate of change of $x(t)$. The autocorrelation functions for two random processes of different rates of change are shown in figure 2.6. Similarly the cross-correlation of two processes is a measure of the change of the degree of relatedness of the processes with time difference. The cross-correlation is defined as:

$$R_{xy}(t_1, t_2) = \mathcal{E}[x(t_1)y(t_2)] = \overline{x(t_1)y(t_2)} \qquad (2.48)$$

From equation (2.36) it follows (by expanding the RHS) that the cross-correlation of $x(t)$ and $y(t)$ at time t_1 (putting $t_2 = t_1$) is:

$$R_{xy}(t_1, t_1) = \rho\sigma_{x_1}\sigma_{y_1} + \overline{x(t_1)}\,\overline{y(t_1)} \qquad (2.49)$$

$$= \rho\sigma_{x_1}\sigma_{y_1} \qquad \text{for zero-mean processes}$$

where σ_{x_1} and σ_{y_1} are the rms values of $x(t)$ and $y(t)$ at time t_1.

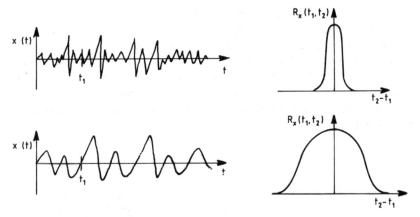

Figure 2.6　Autocorrelation functions of two random processes together with sample time-functions

Since the degree of correlation of x and y decreases as $t_2 - t_1$ increases:

$$\lim_{t_1-t_2\to\infty} R_{xy}(t_1, t_2) = \lim_{t_1-t_2\to\infty} \overline{x(t_1)}\,\overline{y(t_1)} \qquad (2.50)$$

$$= 0 \qquad \text{for zero-mean processes.}$$

2.3.4　*Stationary processes*

If the statistics of a random process are independent of time, the process is termed stationary.

If all the statistics of the process are independent of time then the process is *strictly stationary*. For most purposes it is sufficient if we know that the process is *wide sense stationary*, which requires that the mean and the autocorrelation function are independent of time.

In this case:

$$\overline{x(t)} = \text{constant for all } t \tag{2.51}$$

and we can write:

$$R(t_1, t_2) = R(t') \tag{2.52}$$

where $t' = t_2 - t_1$, that is it is dependent only on the time difference and not on the absolute values of t_2 and t_1.

2.3.5 Ergodic processes

An ergodic process is one in which the complete statistics of the process can be determined from one sample time-function. In other words, time averages are equal to ensemble averages. A necessary condition is that the process is stationary, but the condition of ergodicity is more strict than stationarity.

For a more detailed discussion of ergodicity see Papoulis (1984). In practice, the stationary processes considered in this book are ergodic at least as far as their probability density function, mean and autocorrelation functions are concerned.

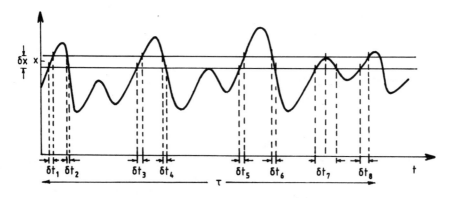

Figure 2.7 Basis of probability density function measurement from time measurements

As an example, the process of calculation of the probability density function from a time rather than an ensemble average is illustrated in figure 2.7. Thus:

$$p(x)\delta x = \lim_{\tau \to \infty} \frac{\sum \delta t_i}{\tau}$$

or:

$$p(x) = \lim_{\tau \to \infty} \frac{\sum \delta t_i}{\tau \delta x} \qquad (2.53)$$

The mean is:

$$\bar{x} = \lim_{\tau \to \infty} \frac{1}{\tau} \int_{-\tau/2}^{\tau/2} x(t)\, dt \qquad (2.54)$$

The mean square value is:

$$\overline{x^2} = \lim_{\tau \to \infty} \frac{1}{\tau} \int_{-\tau/2}^{\tau/2} x^2(t)\, dt \qquad (2.55)$$

and the autocorrelation function is:

$$R_x(t') = \lim_{\tau \to \infty} \frac{1}{\tau} \int_{-\tau/2}^{\tau/2} x(t)x(t - t')\, dt \qquad (2.56)$$

For two jointly ergodic processes, x and y, the cross-correlation function is:

$$R_{xy}(t') = \lim_{\tau \to \infty} \frac{1}{\tau} \int_{-\tau/2}^{\tau/2} x(t)y(t - t')\, dt \qquad (2.57)$$

2.4 The frequency domain

2.4.1 Introduction

In order to determine the response of a linear two-port network (e.g. an amplifier or filter) to a random signal, it is useful to be able to carry out analyses in the frequency rather than the time domain. As we have seen in section 2.3, only averages of measurements on random processes have any utility. Each sample time-function of a random process is unique and in order to calculate the time-function output of a circuit driven by a random signal we would need to know the value of the input signal at all times. However, provided the process is ergodic, each sample time-function of a random process has the same frequency spectrum and the spectrum can be calculated from any one of the time-functions. Knowing the frequency spectrum of the input signal and the frequency response of the network, the spectrum of the output signal may be calculated. It is a property of random signals with a Gaussian pdf that the output of a linear network driven by such a signal also has a Gaussian pdf. If we know the output signal spectrum and that the signal is Gaussian then we shall see that all the other averages of interest (e.g. rms signal level, autocorrelation function) can be calculated.

2.4.2 The Fourier transform

The frequency domain representation of time-function $f(t)$ is given by the Fourier transform:

$$F(f) = \int_{-\infty}^{\infty} f(t)e^{-j2\pi ft}\, dt = \mathcal{F}[f(t)] \tag{2.58}$$

The inverse transform is:

$$f(t) = \int_{-\infty}^{\infty} F(f)e^{j2\pi ft}\, df = \mathcal{F}^{-1}[F(f)] \tag{2.59}$$

where:

$\mathcal{F}[\ \ ]$ is the Fourier transform operator

and:

$\mathcal{F}^{-1}[\ \ ]$ is the inverse Fourier transform operator.

Note that, in this text, time-functions will be written in lower case and the corresponding Fourier transform in upper case.

This transform pair has a number of properties of interest.

1. Symmetry

For real $f(t)$ it is easily seen that:

$$F^*(f) = F(-f) \tag{2.60}$$

where * denotes the complex conjugate.

Therefore:

$$|F(f)| = (F(f)F^*(f))^{1/2} = |F(-f)| \tag{2.61}$$

Thus the modulus of $F(f)$ is an even function of f.

2. Scaling

There is an inverse relationship between the widths of the corresponding functions in the time and frequency domains. Thus:

$$\mathcal{F}[f(\alpha t)] = \frac{1}{|\alpha|} F(f/\alpha) \tag{2.62}$$

from which we can see that short-duration time-functions have wide Fourier transforms and vice versa.

3. Linearity

The Fourier and inverse operators are linear. Thus:

$$\mathcal{F}[a_1 f_1(t) + a_2 f_2(t)] = a_1 F_1(f) + a_2 F_2(f) \tag{2.63}$$

4. Transforms of products

Multiplication in one domain corresponds to convolution in the other. Thus:

$$\mathcal{F}[f_1(t) f_2(t)] = F_1(f) * F_2(f)$$
$$\tag{2.64}$$
$$\mathcal{F}^{-1}[F_1(f) F_2(f)] = f_1(t) * f_2(t)$$

where convolution (*) is defined as:

$$g_1(x) * g_2(x) = \int_{-\infty}^{\infty} g_1(y) g_2(x - y) dy \tag{2.65}$$

Function width increases as a result of convolution. If, for example, $g_1(x)$ and $g_2(x)$ are of finite widths x_1 and x_2 then $g_1(x) * g_2(x)$ has width $x_1 + x_2$.

5. Parseval's theorem

This theorem relates the integrals of products in the two domains. It states:

$$\int_{-\infty}^{\infty} x_1(t) x_2(t) dt = \int_{-\infty}^{\infty} X_1^*(f) X_2(f) df \tag{2.66}$$

for real time-functions $x_1(t)$ and $x_2(t)$.

Four other important properties concerning the Fourier transforms of the differential and integral of time-functions and the effect of time and frequency shifts are the subjects of exercises **2.2** and **2.3** (at the end of the book), and are left to the reader.

2.4.3 Energy and power spectra

A useful parameter of a finite duration voltage or current signal is its normalised energy. This is the energy which the signal would cause to be dissipated in a resistance of 1 ohm. Its units are V^2s or A^2s. Similarly, the normalised power of a continuous signal is the power that the signal would cause to be dissipated in a resistance of 1 ohm. Its units are V^2 or A^2. Dividing the normalised energy or power of a voltage signal, or multiplying the normalised energy or power of a current signal, by the load resistance gives the actual dissipated energy or power.

The normalised energy of a (current or voltage) signal ($f(t)$) of finite duration is obtained from the integral:

$$E = \int_0^\infty f^2(t)\,\mathrm{d}t \tag{2.67}$$

This may be calculated in the frequency domain by making use of Parseval's theorem, noting that $f(t)$ is real. Thus:

$$E = \int_{-\infty}^\infty |F(f)|^2\,\mathrm{d}f \tag{2.68}$$

Since $|F(f)|^2$ is an even function (from equation (2.61)) we can write:

$$E = \int_0^\infty 2|F(f)|^2\,\mathrm{d}f \tag{2.69}$$

Since the energy spectral density is defined as a function $E_s(f)$ where the energy in frequency components lying between f and $f + \mathrm{d}f$ is $E_s(f)\,\mathrm{d}f$, we can see that:

$$E_s(f) = 2|F(f)|^2 \tag{2.70}$$

For a continuous stationary random signal a more useful quantity is the signal power (rate of dissipation of energy), which can be calculated by measuring the energy (E_τ) of the signal within a time τ and dividing by τ. Thus power:

$$P_\tau = \frac{1}{\tau}\int_0^\tau f(t)^2\,\mathrm{d}t$$

$$= \frac{1}{\tau}\int_0^\infty 2|F_\tau(f)|^2\,\mathrm{d}f \tag{2.71}$$

(from Parseval's theorem)

where $F_\tau(f)$ is the Fourier transform of $f_1(t)$
where

$$f_1(t) = f(t) \qquad 0 \leqslant t \leqslant \tau$$
$$= 0 \qquad \text{otherwise}$$

These alternative methods of power calculation are illustrated in figure 2.8.

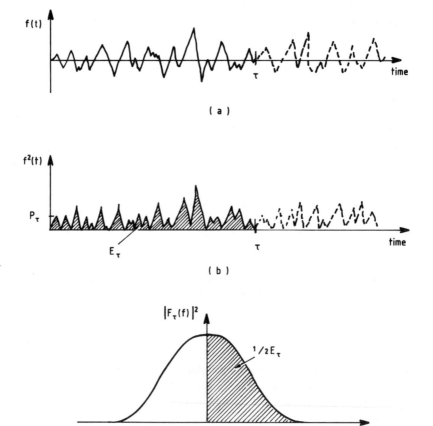

Figure 2.8   Calculation of average signal power over time τ (a) signal (time-function) $f(t)$; (b) signal power $f^2(t)$, energy E_τ (shaded area) and average power P_τ; (c) signal energy E_τ from frequency domain

Note that P_τ is a random quantity since it is the power of one sample time-function. A consistent calculation of power is made by averaging over the ensemble of time-functions. Thus, the average power is:

$$\overline{P_\tau} = \frac{1}{\tau} \int_0^\infty 2\overline{|F_\tau(f)|^2} \, df \qquad (2.72)$$

Note that this is the normalised power of a signal of finite duration τ. In fact

$$F_\tau(f) = \mathcal{F}[f(t)G_\tau(t)]$$

where:

$$G_\tau(t) = 1 \qquad \text{for } 0 < t < \tau$$
$$= 0 \qquad \text{otherwise}$$

This multiplication by $G_\tau(t)$ leads to a convolution by its transform in the frequency domain and therefore a broadening of the calculated transform by approximately $1/\tau$. We eliminate the influence of the finite duration of the time sample by letting τ increase. The average power of $f(t)$ is:

$$\bar{P} = \lim_{\tau \to \infty} \frac{1}{\tau} \int_0^\infty 2\overline{|F_\tau(f)|^2} \, df$$

$$= \int_0^\infty S(f) \, df \qquad (2.73)$$

where:

$$S(f) = \lim_{\tau \to \infty} \frac{2\overline{|F_\tau(f)|^2}}{\tau} \qquad (2.74)$$

$$= \text{the normalised power spectral density or power spectrum}$$

Note that some texts use the bilateral power spectrum:

$$S'(f) = \lim_{\tau \to \infty} \frac{\overline{|F_\tau(f)|^2}}{\tau} \qquad (2.75)$$

as the power spectrum. In this case, $S(f)$ must be integrated over both negative and positive frequencies to obtain the total power. Note

$$S(f) = 2S'(f) \qquad \text{for } f \geq 0$$
$$= 0 \qquad \text{for } f < 0 \qquad (2.76)$$

The relationship between $S(f)$ and $S'(f)$ is shown in figure 2.9.

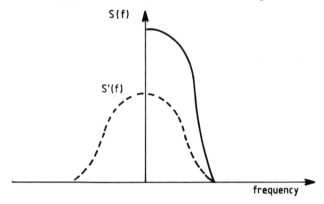

Figure 2.9   Relationships between bilateral $S'(f)$ and unilateral $S(f)$ power spectra

We would expect the autocorrelation function $R(t')$ of a stationary process to be related to the power spectrum since slowly varying signals will be correlated noticeably over longer time durations than will rapidly varying signals (see section 2.3.3). In fact, the Wiener–Khintchine theorem tells us that they are related through the Fourier transform. Thus:

$$S(f) = 2\mathcal{F}[R(t')] \tag{2.77}$$

and, as a consequence of the inverse relationship between the widths of time and corresponding frequency functions, a narrow autocorrelation function (short-duration correlation) is related to a wide power spectrum (rapidly changing signal) and vice versa.

The relationship in equation (2.77) is an alternative (to equation (2.74)) method of calculating the power spectrum.

The relationships between time signals, autocorrelation functions and power spectra are illustrated in figure 2.10.

2.4.4 Linear systems

The frequency domain function linking the input and output of a linear network is called its transfer function. Thus:

$$Y(f) = X_\tau(f)H(f) \tag{2.78}$$

where $X_\tau(f)$ is the Fourier transform of sample of duration τ of input random process $x(t)$, $Y(f)$ is the Fourier transform of the output $y(t)$ and $H(f)$ is the network transfer function.

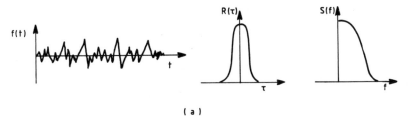

(a)

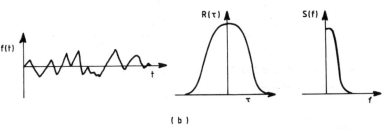

(b)

Figure 2.10   Time function, autocorrelation function and power spectrum:
(a) broad frequency band; (b) narrow frequency band

It follows that the power spectrum of the output is (using equation (2.74)):

$$S_y(f) = \lim_{\tau \to \infty} \frac{2\overline{|X_\tau(f)|^2}}{\tau} |H(f)|^2$$

$$= S_x(f)G(f) \tag{2.79}$$

where:

$$G(f) = |H(f)|^2 \tag{2.80}$$

is the normalised power gain of the network and $S_x(f)$ is the normalised power spectrum of $x(t)$. The relationships between input and output signals, spectra and the network normalised power gain are illustrated in figure 2.11.

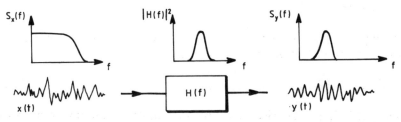

Figure 2.11   Input and output signals, and frequency response of linear network

2.4.5 Cross-power spectra

In section 2.2.8 we saw that the mean square value of the sum of two random variables depended on the degree of correlation between them. This correlation can be expressed in the frequency domain.

Consider the sum of two stationary random processes:

$$z(t) = x(t) + y(t)$$

Taking the Fourier transform gives:

$$Z(f) = X(f) + Y(f)$$

and:

$$|Z(f)|^2 = (X(f) + Y(f))(X^*(f) + Y^*(f))$$

$$= |X(f)|^2 + |Y(f)|^2 + Y(f)X^*(f) + X(f)Y^*(f) \tag{2.81}$$

The bilateral power spectrum is:

$$S_z'(f) = \lim_{\tau \to \infty} \frac{\overline{|Z_\tau(f)|^2}}{\tau}$$

$$= S_x'(f) + S_y'(f) + S_{yx}'(f) + S_{xy}'(f) \tag{2.82}$$

where:

$$S_{yx}'(f) = \lim_{\tau \to \infty} \frac{\overline{Y^*(f)X(f)}}{\tau}$$

and: $\hspace{11cm}$ (2.83)

$$S_{xy}'(f) = \lim_{\tau \to \infty} \frac{\overline{X^*(f)Y(f)}}{\tau}$$

are called the cross-power spectra of $x(t)$ and $y(t)$.

The two spectra are conjugate. Thus:

$$S_{yx}^{*\prime}(f) = S_{xy}'(f) \tag{2.84}$$

and asymmetric, which is the reason for considering the bilateral spectrum. In fact, the real parts of the cross spectra are even functions and the imaginary parts odd. This leads to an alternative form of equation (2.82):

$$S_z'(f) = S_x'(f) + S_y'(f) + 2\,\text{Re}[S_{xy}'(f)] \tag{2.85}$$

Or, since $\text{Re}[S_x'(f)]$ is even, and we can make use of the unilateral spectra:

$$S_z(f) = S_x(f) + S_y(f) + 2\,\text{Re}[S_{xy}(f)] \tag{2.86}$$

Integrating equation (2.85), making use of Parseval's theorem and equation (2.83) gives:

$$\overline{z^2} = \overline{x^2} + \overline{y^2} + 2\overline{xy} \tag{2.87}$$

Comparing these results with those in section 2.2.8, we see that the cross power spectrum is a measure of the degree of correlation of $x(t)$ and $y(t)$ across the frequency spectrum.

There is a link between the cross-correlation function and the cross-power spectrum similar to that between autocorrelation function and power spectrum. Thus:

$$S_{xy}'(f) = \mathcal{F}[R_{xy}(t')] \tag{2.88}$$

If $x(t)$ and $y(t)$ are uncorrelated and zero-mean then:

$$R_{xy}(t') = 0 \quad \text{for all } t'$$

and:

$$S_{xy}'(f) = S_{yx}'(f) = 0 \tag{2.89}$$

and:

$$S_z(f) = S_x(f) + S_y(f) \tag{2.90}$$

The influence of the cross-power spectrum on the spectrum of the sum of two random signals is illustrated in figure 2.12.

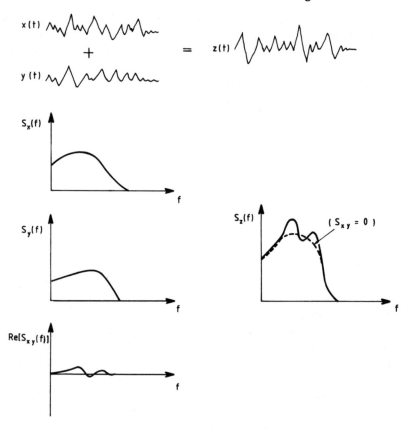

Figure 2.12   Sum of correlated signals

Reference

Papoulis, A. (1984). *Probability, Random Variables, and Stochastic Processes*, 2nd edn, McGraw-Hill, New York.

3 Noise Connected with Layout or Construction

3.1 Introduction

This chapter is concerned with noise arising from a source which is essentially external to the susceptible (low signal level) circuit under consideration – noise which is coupled into the circuit as opposed to the intrinsic noise generated within the circuit elements. This type of noise is strongly linked to circuit layout and construction techniques rather than to the circuit design itself. It is often called interference and may be random or deterministic.

There are many types of noise in this category. The noise source may be electrical in nature and the noise coupled into the susceptible circuit by means of electric, magnetic or electromagnetic fields, or direct connection. This is the concern of the field of electromagnetic compatibility. The source may be mechanical in nature – vibration leading to electrical noise signals by means of the piezoelectric effect, or thermal – leading to random thermal emfs.

Usually the noise source power is high compared with the power of the signals in the circuit of interest and the coupling weak, leading to relatively small but significant noise powers in the susceptible circuit.

It is important to remember, when attempting to reduce this noise, that it is characterised by the requirement of a source plus a pathway (means of coupling) to the susceptible circuit. The noise may be reduced by attention to source, susceptible circuit or pathway.

It is worth emphasising two points. Firstly, that we analyse this type of noise only in order to reduce it to negligible levels, and secondly that the coupling is often complex and unclear.

For these reasons, a degree of approximation is often used which would be unacceptable elsewhere. The interest is not whether a noise level is 1.0 or 1.5 μV compared with a 10 μV signal but in making sure it is not 10 μV or greater.

31

3.2 Field-coupled noise

3.2.1 Introduction

Of all the types of noise of electrical origin, field-coupled noise is probably one of the most difficult to compute. In general, any time-varying current in a circuit has associated with it both an electric and a magnetic field which radiate from the circuit and which may be calculated using Maxwell's equations. Indeed, the theoretical basis for field coupled noise is just this set of equations which may, in theory, be used to calculate the current in any part of the circuit, given the source current characteristics and the construction of the circuit and any other circuit coupled to it. In practice, the boundary conditions, even in simple cases, are extremely complex and gross simplifications are required in many cases in order to obtain a manageable problem before attempting to find a solution.

Over distances which are small compared with $\lambda/(2\pi)$, where λ is the free-space wavelength of the radiated field, the coupling between circuits may be calculated by considering electric and magnetic fields separately – that is, by considering coupling by interconductor mutual capacity and mutual inductance. Since the wavelength at 300 MHz is 1 metre, and the inter-conductor distances in an electronic instrument are usually a few centimetres or less, this condition is usually met.

At distances greater than $\lambda/(2\pi)$ from a small source, the radiated electromagnetic field is dominant and coupling is by means of this field. It is usually most important for coupling radio frequency signals from sources outside the instrument.

These different fields are discussed more fully in chapter 9.

3.2.2 Electromagnetic coupling

Electromagnetic radiation is emitted from any circuit in which the current is varying. For example, the magnitude of the electric field at distance r ($>> \lambda/(2\pi)$) from a short straight wire of length l ($<< \lambda/(2\pi)$) carrying a current I at frequency f (figure 3.1(a)) is (Corson and Lorrain, 1962):

$$|E_e| = \frac{Ilf\mu_0 \sin(\theta)}{2r} \quad \text{V/m}$$

$$= \frac{0.2\pi Ilf\sin(\theta)}{r} \times 10^{-6} \quad \text{V/m} \tag{3.1}$$

and that from the same current flowing in a small (diameter $<< \lambda/(2\pi)$) loop of area A (figure 3.1(b)) is:

$$|E_m| = \frac{\mu_0^{3/2}\varepsilon_0^{1/2}\pi IAf^2 \sin(\theta)}{r} \quad \text{V/m}$$

$$= \frac{0.1316\,IAf^2 \sin(\theta)}{r} \times 10^{-13} \quad \text{V/m} \tag{3.2}$$

Both these equations assume isolated conductors and in practice the presence of other structures and, in particular, conductors, will alter the field. In addition, the current in an open-ended straight wire drops to zero at the end, and the assumption of uniform current density is an approximation. However, these equations serve to give order-of-magnitude figures and to illustrate the dependence on radiator dimensions, frequency and distance.

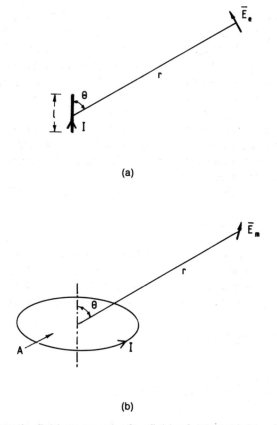

(a)

(b)

Figure 3.1   Electric field vectors in far field of (a) straight wire (electric dipole) and (b) loop (magnetic) dipole radiators

The dependence on frequency means that only at high frequency is there significant radiation. The field at 10 metres ($\theta = 90°$) from a straight wire radiator of length 10 cm and a loop radiator of area 10 cm^2, both carrying currents of 10 mA at a number of frequencies, is shown in table 3.1.

Table 3.1 Electric field from electric ($|E_e|$ and magnetic dipole ($|E_m|$ radiators

Frequency (Hz)	$|E_e|$ (V/m)	$|E_m|$ (V/m)
10	0.628×10^{-9}	0.132×10^{-17}
10^4	0.628×10^{-6}	0.132×10^{-11}
10^7	0.628×10^{-3}	0.132×10^{-5}

Sources of electromagnetic radiation include radio transmitters, medical diathermy apparatus, arcs in brush motors, spark plugs, switches and thermostats, and circuits operating at high frequencies (such as computers). The radiation, from sources other than radio transmitters (figure 3.2(a)) where the radiation emission is deliberate, is from (sometimes spurious) high frequency currents flowing in conductors within the device and connecting cables (figure 3.2(b)) (for example, power-line cable).

The radiated interference from high frequency electronic circuits is electric dipole radiation from unscreened cables and magnetic dipole radiation from current loops (figure 3.2(c)).

The radiation may be picked up by any conductor in a circuit acting as an antenna, and clearly the most susceptible circuits are those in which the signal level is low. It is important to note that this problem does not affect only high frequency circuits. A demodulation of a radio frequency signal can take place at any non-linearity in a circuit, leading to the modulation signal (often having a significant audio frequency component) appearing in the susceptible circuit. The non-linearity may arise as a result of overload of the input stage of a low-level low frequency amplifier by a high-level radio frequency signal – the electric field strength in the vicinity of radio transmitters can be several volts per metre. On the other hand, even circuits operating well below limiting levels can have a sufficient degree of non-linearity to lead to significant modulation signal noise.

The means of reduction of this type of noise will be discussed in detail in chapter 9 and is covered only briefly here.

The radiation may be reduced at source (except for radio transmitters) by reducing high frequency currents in the source circuits. High frequencies are associated with short rise and fall times in digital circuits, and limiting these to that required for correct circuit operation will limit the high frequencies

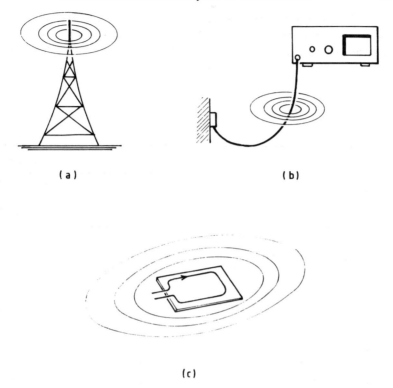

(a)

(b)

(c)

Figure 3.2   Electromagnetic radiation at rf: (a) broadcast transmitter; (b) unshielded cable; (c) current loop on PCB

within the circuit and the radiated field. Reducing the length of wires and the area of loops carrying the current will also help. Since electromagnetic fields are strongly reflected from conductors and attenuated within them, shielding the source and/or susceptible circuit by enclosing these, as far as possible, in a conductive casing will reduce the field strength. Low-pass filters may be used to reduce spurious high frequency signals on power leads and to reduce noise signal levels in susceptible circuits. Note that, in the latter case, the filters should preferably be placed before any semiconductor device in order to avoid the non-linearity demodulation problem.

These interference reduction techniques are illustrated in figure 3.3.

3.2.3   *Magnetic (mutual inductance) coupling*

A time-varying magnetic field crossing the plane of a conductor loop induces an emf in the loop of:

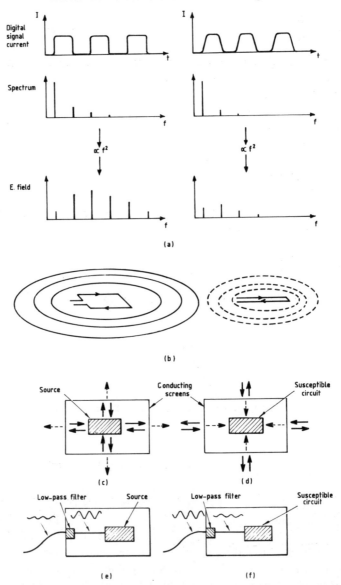

Figure 3.3   Methods of limiting em-coupled noise: (a) increasing rise and fall times of digital signal current flowing in loop reduces harmonics and radiated field; (b) decreasing area of radiating current loop reduces field; (c) reflection from, and attenuation in, a conducting screen reduces radiation from source of em field; (d) reflections from, and attenuation in, a conducting screen reduces em field at susceptible circuit; (e) low-pass filter reduces radio frequency signal on supply cable to signal source; (f) low-pass filter reduces radio frequency signal induced on supply, or signal cable acting as antenna

$$v = -\frac{d\phi}{dt} \qquad (3.3)$$

where the flux ϕ may be calculated from the integral of the magnetic induction or flux density vector $\bar{B}$ (Webers/m^2) over the area of the loop (figure 3.4). That is:

$$\phi = \int_A \bar{B} \cdot d\bar{A}$$

$$= \int_A B_n dA \qquad (3.4)$$

where B_n is the component of $\bar{B}$ perpendicular to the plane of the loop. Combining this with (3.3) gives:

$$v = -A\frac{dB_A}{dt} \qquad (3.5)$$

where B_A is the average of B_n over the loop area.

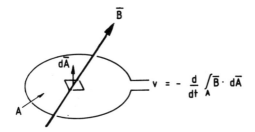

Figure 3.4   Calculation of induced emf

The magnetic field is usually associated with current flowing in another circuit and if ϕ_{21} is the flux in secondary circuit 2 resulting from a current i_1 in primary circuit 1, then they are related by the mutual inductance M_{21} between the circuits (figure 3.5). That is:

$$M_{21} = \frac{\phi_{21}}{i_1} \qquad (3.6)$$

and from (3.3):

$$v_2 = -M_{21}\frac{di_1}{dt} \qquad (3.7)$$

That is, the emf induced in the secondary circuit is proportional to the rate of change of the current flowing in the primary circuit.

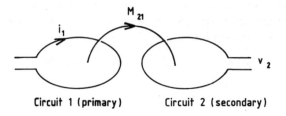

Figure 3.5　Mutual inductance of two circuits

It is often more convenient to perform analyses in the frequency domain, in which case the relationship between the Fourier components of v_2 and i_1 at frequency f is (see exercise **2.2** at end of book):

$$V_2(f) = j2\pi f M_{21} I_1(f) \tag{3.8}$$

or:

$$|V_2(f)| = 2\pi f M_{21} |I_1(f)| \tag{3.9}$$

If the current i_1 has spectral density $S_{ni}(f)$ A^2/Hz, then the spectral density of v_2 is:

$$S_v(f) = (2\pi f M_{21})^2 S_{ni}(f) \quad \text{V}^2/\text{Hz} \tag{3.10}$$

and the normalised noise power is:

$$\overline{v^2} = \int_0^\infty S_v(f)\mathrm{d}f$$

$$= (2\pi M_{21})^2 \int_0^\infty f^2 S_{ni}(f)\mathrm{d}f \quad \text{V}^2 \tag{3.11}$$

Common sources of these magnetic fields are power-line transformers and electric motors where the fields and induced emfs are at line frequency (50–60 Hertz), although the field from brush motors will contain some higher frequency components as a result of arcing.

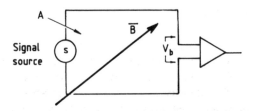

Figure 3.6　Magnetic induction of noise in amplifier input circuit

Noise coupling into the input circuit of an amplifier is shown in figure 3.6. For example, if the stray field of a power transformer has an average flux density normal to the loop formed by the input circuit of:

$$B_A = B_0 \cos(2\pi f t) \tag{3.12}$$

then, from (3.5), the induced noise emf is:

$$v_b = 2\pi f B_0 A \sin(2\pi f t) \tag{3.13}$$

If $B_0 = 1 \, \mathrm{mWb/m^2}$, $f = 50 \, \mathrm{Hz}$, $A = 1 \, \mathrm{cm^2}$, then the amplitude of v_b is:

$$v_{b0} = 2\pi f B_0 A$$

$$= 31.4 \quad \mu V$$

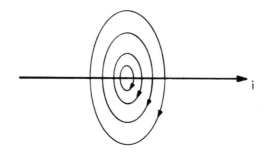

Figure 3.7   Magnetic field from single conductor carrying current *i*

The magnetic field may be generated by the current flowing through a single wire as shown in figure 3.7. The field at a distance *r* from a long straight wire carrying a current *i* is (Corson and Lorrain, 1962):

$$B = \frac{\mu_0 i}{2\pi r}$$

$$= 2 \times 10^{-7} i/r \quad \mathrm{Wb/m^2} \tag{3.14}$$

For example, the flux density at 1 cm from a wire carrying a current of 1 A is $20 \, \mu \mathrm{Wb/m^2}$. This would lead to an induced noise voltage of 1/50th of that in the example above if the susceptible circuit loop was positioned such that this was the average flux density over the 1 cm² loop area.

Note that we have assumed that the conductor(s) carrying the return current are sufficiently distant that their magnetic field is negligible.

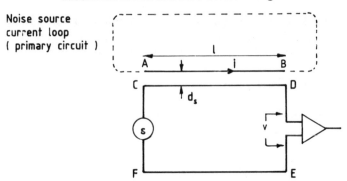

Figure 3.8   Noise-coupling via mutual inductance

Probably a more common situation is that shown in figure 3.8 where the secondary circuit loop area is much larger, and the flux density is much higher in the vicinity of that part of the secondary loop close to the noise source conductor than at the far side of the loop. In this situation, the induced emf may be calculated using the mutual inductance of the interference source (primary) circuit and the susceptible (secondary) circuit and, if the conductors in each circuit run close together over a significant length and are far apart elsewhere, then the mutual inductance of the two circuits may be calculated using an approximate formula involving only the dimensions and separation of the neighbouring conductors.

For example, the mutual inductance of two circuits where the close conductors have circular cross-sections, and are non-magnetic with wires running parallel over length l is (Giacoletto, 1977):

$$M = \frac{\mu_0 l}{2\pi} \left(\ln \left(2l/d_s \right) - 1 \right) \quad \text{H} \tag{3.15}$$

$$= 0.2l \left(\ln(2l/d_s) - 1 \right) \quad \mu\text{H} \tag{3.16}$$

where d_s is their separation (centre-to-centre) and $l \gg d_s$. The mutual inductance is independent of conductor diameter.

Note that this is often loosely described as the mutual inductance of two parallel wires.

For example in the circuit of figure 3.8, if l is 10 cm, d_s is 2 mm, then M is 0.072 µH.

If conductor AB is part of a digital circuit and carries pulse currents with maximum rates of change of current of 10^6 A/s (for example, 10 mA in 10 ns) then the maximum induced emf in the secondary circuit CDEF is, from equation (3.7), 72 mV as shown in figure 3.9(a).

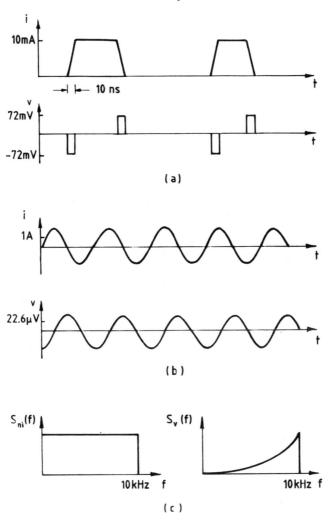

Figure 3.9 Noise coupled through mutual inductance: (a) digital signal current; (b) power supply current; (c) broadband noise current. $S_v(f)$ is the spectrum of v

If AB is a power-line, carrying a current of 1 A rms at 50 Hz, then the induced emf is, from (3.9), 22.6 µV (figure 3.9(b)).

If the current i is broadband noise with a constant spectral density $10^{-6} A^2/Hz$ over frequency range 0–10 kHz (figure 3.9(c)) then the induced rms noise emf may be calculated using (3.11), as:

$$v_{rms} = 2\pi M \left(\int f^2 S_{ni}(f) df \right)^{1/2}$$

$$= 2\pi \times 0.072 \times 10^{-6} \left(\int_0^{10^4} 10^{-6} f^2 df \right)^{1/2} \quad V$$

$$= 2\pi \times 0.072 \times 10^{-9} \left[f^3/3 \right]_{f=10^4}^{1/2} \quad V$$

$$= 0.26 \quad mV$$

The mutual inductance of two circuits is affected by the proximity of a conductor in a third circuit, particularly if this circuit is low impedance, since the current flowing in this third circuit as a result of an induced emf affects the flux in the other circuits. For example, if the conductors in the situation described above run parallel to a conducting ground plane as shown in figure 3.10, then the mutual inductance between the circuits is (Baker *et al.*, 1970):

$$M \simeq \frac{\mu_0 l}{4\pi} \ln(1 + 4(h/d_s)^2) \quad H \tag{3.17}$$

$$= 0.1 l \ln(1 + 4(h/d_s)^2) \quad \mu H \tag{3.18}$$

where l is the length of the wire, h ($<< l$) is the height of the wire centres above the ground plane and the wire diameters are small compared with h and d_s.

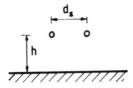

Figure 3.10   Thin parallel wires running parallel to ground-plane

For example, if the wires of the previous example are 2 mm above a ground plane, then their mutual inductance is reduced from 0.072 μH to 0.016 μH.

It is possible, therefore, to reduce mutual inductance and hence magnetic field-coupled noise by running conductors close to a ground plane.

Again, methods of reducing noise coupled by magnetic fields will be discussed in detail in chapter 9 and therefore only briefly here. The methods mentioned are illustrated in figure 3.11.

Clearly, low signal level circuits should be sited away from sources of time-varying magnetic fields. However, if sufficient spacing is not possible, then there are a number of precautions and noise reduction techniques which can be adopted.

Inductively coupled noise from transformers may be reduced at the source, by using transformers with low magnetic leakage (for example, toroidal transformers) and by screening – enclosing the transformers within a high permeability container, to reduce the flux leakage outside the screen. The susceptible circuit may be screened in a similar fashion in order to reduce the flux within the screen. A common high permeability screening material is mu-metal. A disadvantage of high permeability material is that it tends to saturate at fairly low magnetic field strengths so that at high field strengths it becomes ineffective as a shield. This problem can be alleviated by using two shields – an outer shield of lower permeability magnetic material but with a high saturation level and an inner shield of high permeability material. The outer shield will then reduce the magnetic field strength to a level below the saturation level of the high permeability shield.

Permeability also decreases with increasing frequency, but a high (radio) frequency magnetic field can be reduced by the use of a conducting, non-magnetic shield since the induced eddy currents in the shield generate a magnetic field in opposition to the field producing it.

The area of the susceptible circuit should be reduced to a minimum (see equation (3.4)) in order to reduce the flux. Use of twisted paired wires is of benefit here since not only are the leads close together, but the twists lead to areas of alternating sign leading to partial cancellation of the induced emfs.

The orientation of the coil can be changed such that the time-varying magnetic induction vector is parallel to the plane of the coil.

The conductors in low noise circuits should not run close to conductors carrying time-varying high-level currents and reduced coupling may be achieved by running the leads close to a ground plane.

Rise and fall times in potentially interfering circuits should be as long as is tolerable in order to reduce the noise coupling (see equation (3.7)).

3.2.4 Electric (capacitative) coupling

There is spurious electrical coupling between conductors in electrical circuits by means of the electric field between conductors, or their mutual capacitance (figure 3.12). The current flowing through this mutual capacitance C is proportional to the rate of change of the potential difference across the capacitor. That is:

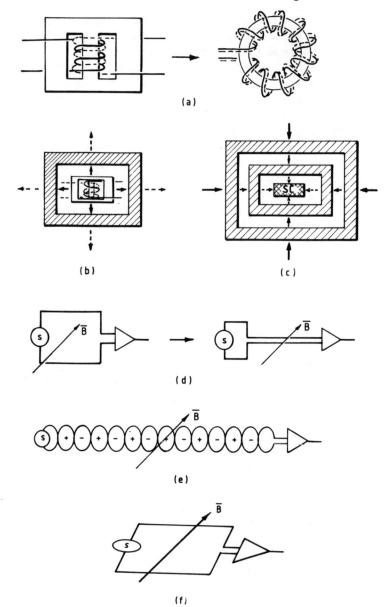

(a)

(b)

(c)

(d)

(e)

(f)

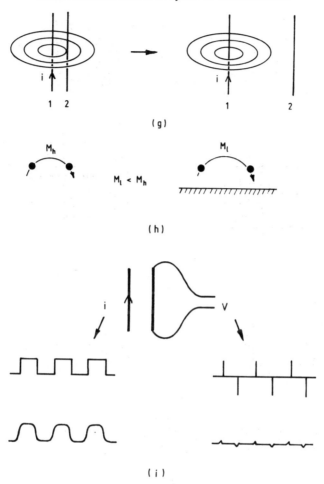

Figure 3.11 Methods of reducing inductively coupled noise: (a) change to low leakage transformer; (b) shielding source – reduction of magnetic field using magnetic material at low frequencies and non-magnetic conductor at high frequencies; (c) shielding susceptible circuit (SC) at low frequencies using medium permeability outer, and high permeability inner, screens; (d) flux reduction by reduction of loop area; (e) net flux reduction in twisted pair by alternating sign of area; (f) change of relative orientation such that magnetic induction vector is parallel to plane of loop; (g) increase separation of noise source (1) and susceptible circuit (2) conductors; (h) reduction of mutual inductance using ground plane; (i) reduction of di/dt in noise source conductor

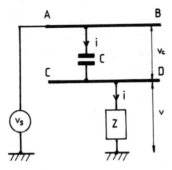

Figure 3.12 Coupling of noise from conductor AB to CD by means of mutual capacitance C

$$i = C\frac{dv_c}{dt} \tag{3.19}$$

or in the frequency domain (see exercise **2.2** at end of book):

$$I = j2\pi f C v_c \tag{3.20}$$

Since we normally know v_s, the potential on the noise source conductor with respect to local ground, it is convenient to express the noise in the susceptible circuit in terms of this. The mutual capacitance and the impedance (to ground) of the susceptible circuit form a voltage divider and, in the frequency domain:

$$v/v_s = Z/(Z - j/(2\pi fC))$$

$$= (1 + (j2\pi fCZ)^{-1})^{-1} \tag{3.21}$$

Often the reactance (X_c) of C is very much greater than $|Z|$, in which case:

$$v/v_s \simeq j2\pi\, fCZ \tag{3.22}$$

Note that this condition is equivalent to saying that the current through C is determined by the reactance of C and v_s or $v_c \simeq v_s$. If Z is resistive $(Z = R)$, then, under these conditions, the voltage across R in the time domain is, from (3.19):

$$v = iR$$

$$= CR\frac{dv_s}{dt} \tag{3.23}$$

Note that if these conditions ($Z = R$ and $X_c \gg Z$) are not met, then v can be obtained by inverse-Fourier-transforming V obtained from (3.21).

If v_s has a spectral density $S_{vs}(f)$, then the noise voltage v has spectral density (from equation (3.21)):

$$S_{nv}(f) = S_{vs}(f)|1 + (j2\pi f CZ)^{-1}|^{-2} \tag{3.24}$$

or:

$$S_{nv}(f) = S_{vs}(f)|j2\pi f CZ|^2 \tag{3.25}$$

if $X_c \gg |Z|$.

The second factor on the right-hand side of (3.24) and (3.25) is the normalised power frequency response of the filter formed by C and Z ($|v/v_s|^2$, from equation (3.21)).

In order to get an idea of the magnitude of the effect, we again consider two parallel conductors. The mutual capacitance of two parallel circular conductors of diameter d, axis separation d_s and length l ($l \gg d_s, l \gg d$) is (Giacoletto, 1977):

$$C = \frac{\pi \varepsilon l}{\cosh^{-1}\left(\dfrac{d_s}{d}\right)} \tag{3.26}$$

$$= \frac{\pi \varepsilon l}{\ln\left(\dfrac{2d_s}{d}\right)} \quad \text{(for } d_s \gg d, \text{ within 5 per cent if } d_s/d > 2) \tag{3.27}$$

where ε is the permittivity of the surrounding medium.

For example, if the conductors are in air ($\varepsilon = \varepsilon_0 = 8.85 \times 10^{-12}$ F/m) and if $d_s = 2$ mm, $d = 1$ mm and $l = 10$ cm, then C is 2.0 pF.

As an example, consider the circuit shown in figure 3.13, where noise is coupled from lead AB through mutual capacitance C ($= 2.0$ pF) into a circuit with resistance R ($= 10$ kΩ) to ground.

If v_s is a digital signal of maximum rate of change (dv_s/dt) of 2×10^6 (2 volt in 1 μs) then, from equation (3.19) (taking $v_c \simeq v_s$), $i = 4$ μA and $v = 40$ mV (figure 3.14). If the amplifier shown is acting as a buffer between an analogue signal source and a 10-bit ADC with voltage range 0–10 volt, then the resolution of the ADC is $10/1024 = 0.0098$ volt (9.8 mV) and the 40 mV noise represents an appreciable loss of performance amounting to an error of over 2 bits.

If AB is a power supply lead such that v_s has an rms value of 250 volts at 50 Hz then, from (3.22), v has an rms value of 1.6 mV.

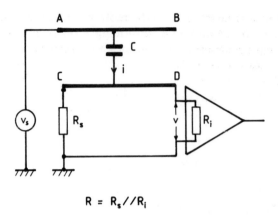

$$R = R_s // R_i$$

Figure 3.13   Noise coupled into amplifier input circuit

If v_s is a broadband noise signal, then the rms level of v can be calculated from (3.24) or (3.25). If the spectral density is constant at 10^{-6} V^2/Hz over a frequency range 0–10 kHz and zero outside this range then, firstly, we establish that equation (3.25) may be used since $X_c \gg R$ at the highest frequency $(1/(2\pi f C) = 8\,M\Omega$ at 10 kHz), and then calculate v_{rms} as:

$$v_{rms} = 2\pi CR\left(\int_0^{10^4} f^2 S_{vs}(f)df\right)^{1/2}$$

$$= 2\pi \times 2 \times 10^{-12} \times 10^4 \times 10^{-3} \times \left[f^3/3\right]_{f=10^4}^{1/2} \quad V$$

$$= 72.6 \quad \mu V$$

It is clear that, since stray capacitances are normally small and therefore have high reactances, this type of noise coupling will normally be a problem in high impedance, low-level circuits.

However, it is important not to be misled by this rule of thumb. For example, it is tempting to treat the output impedance of operational amplifiers with feedback as extremely low and therefore not susceptible to capacitatively coupled interference. However, the output impedance of low frequency amplifiers can be quite high at high frequencies as a result of the reduction in open loop gain as the frequency increases, with output impedances of 10–100 Ω being not uncommon. This can allow capacitative coupling from high frequency parts of the circuit, for example, leads carrying logic signals as shown in figure 3.15.

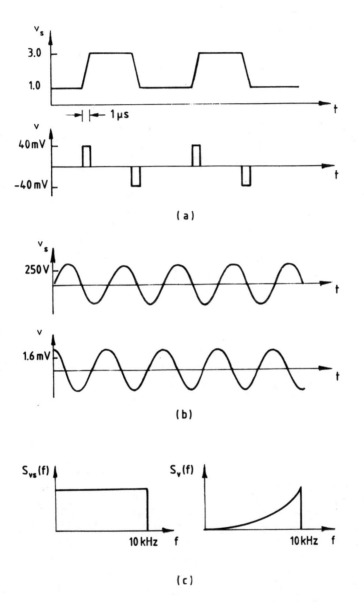

Figure 3.14 Noise coupled through mutual capacitance from source: (a) digital signal; (b) power supply; (c) broadband noise

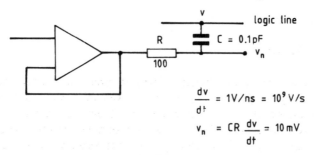

Figure 3.15   Noise from fast switching signals capacitatively coupled to amplifier output lead

As with magnetic field coupling, the presence of a nearby third conductor affects the electric field and therefore the mutual capacitance of the two conductors. For example, the mutual capacitance between two parallel circular cross-section wires at height h above a ground plane is (Baker *et al.*, 1970):

$$C = \frac{\pi\varepsilon l \ln(1 + 4(h/d_s)^2)}{(\ln(4h/d))^2 - (0.5\ln(1 + 4(h/d_s)^2))^2} \qquad (3.28)$$

for $h \gg d$, $d_s \gg d$, $l \gg d_s$ and $l \gg h$.

It shows that the mutual capacitance between two wires reduces as the separation between them and the ground plane reduces.

It is usually difficult to estimate accurately the mutual capacitance of conductors since it is affected not only by the proximity of conductors but also insulators since their permittivity differs from that of free space. There is no simple expression, for example, for the mutual capacitance of strip conductors on the surface of an insulator (as in a PCB).

Again, the means of reducing electric field-coupled noise will be discussed in detail in chapter 9. Briefly (and illustrated in figure 3.16), the mutual capacitance between conductors can be reduced by increasing conductor separation and by the close proximity of a ground plane, and by placing conductors connected to local ground between the conductors carrying potentially interfering signals and the conductors carrying low-level signals. The interference problem can be reduced by the use of screened leads and by the use of twisted pairs together with differential amplifiers, since the capacitances between the interference source wire and the two low-level signal wires are similar, and the interference signal appears as a common mode signal.

Again, the interference from digital circuits can be reduced by limiting pulse rise and fall slopes.

3.3 Conducted noise

3.3.1 Introduction

A direct connection from a noise-generating circuit to a noise-susceptible circuit is clearly the most straightforward way of noise coupling and, it might be thought, the most easily avoided. However, in most electronic systems there are, of necessity, connections between susceptible circuits – usually at the input of an analogue circuit sub-system – and noise-generating, and therefore, potentially interfering circuits. These noise source circuits may be power supplies, the line supply, digital and switching circuits or high power analogue circuits. The noise path is often the local ground to which these circuits and the susceptible, low signal level, analogue circuits are connected.

3.3.2 Common signal path

Noise can be coupled into a susceptible circuit as a result of the 'wanted' signal and a noise signal sharing a connection. This occurs most commonly, but not exclusively, in the ground connection, and a typical case is shown in figure 3.17(a). This is equivalent to the circuit in 3.17(b). The ground connection AD carries supply current, not only to the amplifier shown, but also to other parts of the circuit. The connection between B and C has a non-zero impedance Z_c leading to a voltage drop V_c across BC which appears in series with the signal at the amplifier input. Two factors determine the nature and magnitude of the noise voltage V_c – the current I_c and the impedance Z_c. We assume in our analysis that the input impedance of the amplifier plus Z_s is very much greater than Z_c.

If I_c is simply a constant supply current I_{DC} then V_c is constant $(I_{DC}R_c)$ and may be treated as part of the amplifier input offset voltage, or eliminated by high pass filtering if the source has no required steady component. However, the current I_c will normally have a small but sometimes significant noise component I_n. Thus:

$$I_c = I_{DC} + I_n \tag{3.29}$$

and V_c has a noise component:

$$V_n = I_n Z_c \tag{3.30}$$

where:

$$|Z_c| = (R_c^2 + (2\pi f L_c)^2)^{1/2} \tag{3.31}$$

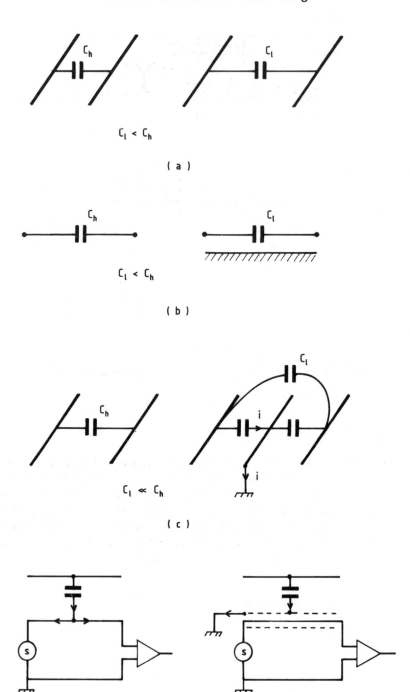

(a)

$C_l < C_h$

(b)

$C_l < C_h$

(c)

$C_l \ll C_h$

(d)

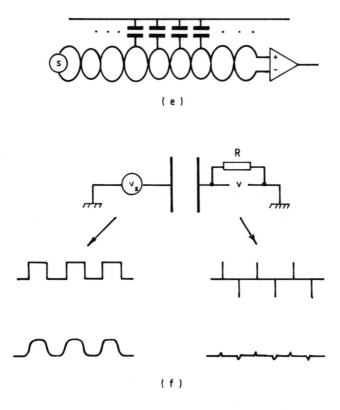

(e)

(f)

Figure 3.16 Method of reducing capacitatively coupled noise: (a) increase conductor separation; (b) reduction of mutual capacitance using ground plane; (c) use of grounded conductor (partial screening); (d) use of screened cable; (e) conversion to common mode noise using twisted pair connection, floating (or balanced source) and differential amplifier; (f) reduction of dv_s/dt on noise source conductor

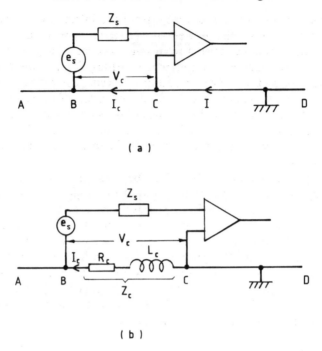

(a)

(b)

Figure 3.17   Noise from common signal path: (a) amplifier input circuit; (b) equivalent circuit including parasitic components in ground line

Possible sources of the noise current are shown in figure 3.18.

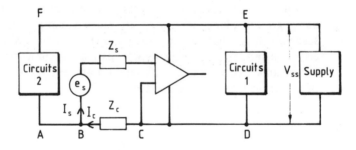

Figure 3.18   Sources of ground-line noise current

Noise superimposed on the supply voltage V_{ss} will give rise to noise in the supply current. This can be generated within the power supply and will have a component of random noise arising from the intrinsic noise within the circuit components of the power supply, a component at power-line frequency and harmonics arising from imperfect smoothing of the output of the supply rectifier, and a component arising from power-line borne interference (see section 3.3.4). In addition, varying current demand by all the circuits supplied will cause sympathetic variation of the supply voltage as a result of the finite output impedance of the supply and the impedances of the connecting leads. In addition, the current through Z_c will be modulated by the varying current demand of circuits 2 in figure 3.18 even if V_{ss} is constant.

Referring to equation (3.30), it can be seen that the other quantity determining the noise voltage at the input of the amplifier is the conductor impedance Z_c. The impedance has a resistive and inductive component, and the inductive reactance will often exceed the resistance even at quite low frequencies. The resistance of a conductor increases as the frequency increases as a result of the progressive concentration of current density near the surface – the skin effect (see section 9.2.3). It is not generally appreciated that this effect is also often significant at quite low frequencies.

The skin depth – the depth at which the current density has dropped to e^{-1} of its value at the surface is (Hayt, 1981):

$$\delta = (\pi f \mu \sigma)^{-1/2} \tag{3.32}$$

At sufficiently high frequencies, where the current density at the centre of a conductor is a negligible fraction of that at the surface – that is to say, δ is small compared with the smallest dimension of the conductor – the resistance can be calculated by considering that the current density is uniform and concentrated only within a distance δ from the surface and the AC resistance may be easily calculated. When δ is comparable with the conductor width, the expressions for AC resistance are complicated. Values are tabulated for regular cross-section geometries (Giacoletto, 1977).

The AC resistance of a circular cross-section conductor of length l and diameter d where $d \gg \delta$ is (Baker *et al.*, 1970):

$$R_{AC} = l/(\pi d \sigma \delta) \tag{3.33}$$

whereas the DC resistance is:

$$R_{DC} = 4l/(\pi d^2 \sigma) \tag{3.34}$$

The ratio of the AC and DC resistances is:

$$R_{AC}/R_{DC} = d/(4\delta) \tag{3.35}$$

For copper ($\sigma = 5.82 \times 10^7 \text{ S m}^{-1}$) and:

$$\delta = 0.066 f^{-1/2} \quad \text{m} \tag{3.36}$$

$$R_{AC}/R_{DC} = 3.79 d f^{1/2} \tag{3.37}$$

For example, the resistance of 10 cm of 1 mm diameter copper wire is 2.19 mΩ for DC and 26.2 mΩ at 10 MHz.

The proximity of a conductor carrying the return current – the pair forming a transmission line – affects the current distribution and therefore the AC resistance. In this case, the current is concentrated on the facing surfaces (Terman, 1955; Hayt, 1981).

The presence of a metal/metal contact such as a plug–socket connection or a switch contact can increase the path resistance significantly. Typical contact resistances are in the range 2–20 mΩ as quoted by manufacturers, but these may increase significantly if the contact becomes dirty.

The inductance of a conductor has two components – that resulting from the magnetic field within the conductor – its internal self-inductance – and that resulting from the external field – the external self-inductance. The latter component usually dominates. For example, the internal self-inductance per unit length of a cylindrical non-magnetic conductor is 0.05 µH/m (independent of diameter) at low frequencies and decreases, as a result of the skin effect, as the frequency increases (Hayt, 1981).

The external self-inductance of a straight, non-magnetic cylindrical conductor of circular cross-section, diameter d and length l ($l \gg d$) is (Giacoletto, 1977):

$$L \simeq 0.2 l \left(\ln(4l/d) - 1 \right) \quad \text{µH} \tag{3.38}$$

and for a rectangular cross-section with width w and thickness t:

$$L \simeq 0.2 l \left(\ln(2l/(w + t)) + 0.5 \right) \quad \text{µH} \tag{3.39}$$

For example, the 10 cm of 1 mm diameter copper wire ($R_{DC} = 2.19 \text{ mΩ}$) discussed above has an inductance of 0.1 µH, which has a reactance of 6.3 mΩ at 10 kHz and 6.3 Ω at 10 MHz.

At power-line frequency (50–60 Hz), when the reactance of L_c compared with R_c and the skin effect are both negligible, a noise current I_n of 1 mA will lead to a noise voltage V_n of 2.19 µV. At 10 MHz, the same current leads to a noise voltage of 6.3 mV.

If the noise current has a normalised spectral density of $S_{ni}(f)$ then:

$$\overline{v_c^2} = \int S_{ni}(f)(R_c^2 + (2\pi f L_c)^2) \mathrm{d}f \tag{3.40}$$

Remember that, in general, R_c is a function of frequency, although it may be treated as constant, provided that the bandwidth of $S_{ni}(f)$ is sufficiently small so that δ does not change significantly.

It should be noted that the use of inductances calculated from the above formulae in order to calculate conductor impedance, and from that the noise voltage, is likely to lead to only very approximate figures. This arises from the nature of inductance. Strictly, only a closed loop may be said to have an inductance (Weber, 1965). The inductance of a closed loop may be calculated from a consideration of its component conductors, but the total inductance requires not only the self-inductances of the component conductors but also the mutual inductances of all the pairs of conductor components and pairs formed from these component conductors and conductors in other circuit loops (Ruehli, 1972). A closer approximation is shown in figure 3.19. Clearly these considerations properly belong in section 3.2.3. However, since it is important to consider them together with connection resistance and self-inductance, we illustrate their effect here.

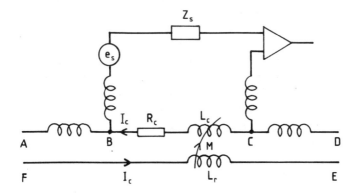

Figure 3.19   The amplifier input circuit of figure 3.17(a) showing spurious inductive components. Note that each pair of conductors for which the self-inductance is shown also has a mutual inductance – only that between BC and FE is shown

As an example, if the return current conductor EF in figure 3.19 runs close to BC and we neglect all the other mutual inductances then:

$$V_c = I_c(R_c + j\,2\pi f L_c - j\,2\pi f M)$$

$$= I_c(R_c + j\,2\pi f L_e) \tag{3.41}$$

where:

$$L_e = L_c - M \tag{3.42}$$

From (3.16) and (3.38), L_e, for two parallel circular conductors of diameter d, and with centres separated by d_s, running parallel over distance l ($l \gg d_s$) is:

$$L_e = 0.2l \ln(2d_s/d) \tag{3.43}$$

If $d_s = 5\,\mathrm{mm}$, the equivalent inductance is $0.046\,\mu\mathrm{H}$ compared with $0.1\,\mu\mathrm{H}$ when the conductor BC was considered independently.

Note that the assumption has been made that the impedance of the capacitance between AD and EF is negligible compared with Z_c in estimating V_c. It is easily shown that this is the case for the frequencies considered.

Common path noise may be reduced by a number of design strategems.

If the source can be grounded at the amplifier input, as shown in figure 3.20(a), then the voltage V_c no longer appears in the input circuit. Alternatively, if the source has to be grounded at B, a differential amplifier may be used, as shown in figure 3.20(b). Note that this requires a circuit with good common mode rejection and high input impedance since the source is unbalanced.

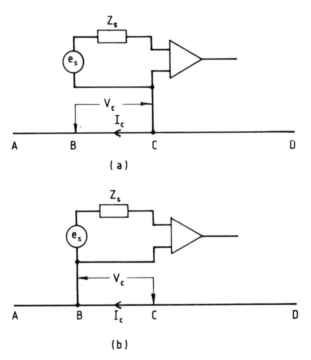

Figure 3.20 Methods of reducing common path noise: (a) single common connection at amplifier input; (b) differential amplifier with single common connection at source

With reference to figure 3.18, connecting the supply to circuit 2 such that its supply (or signal) currents do not flow through BC (figure 3.21) reduces V_c by reducing the noise current flowing through Z_c.

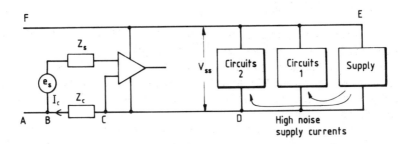

Figure 3.21   Re-configuration of circuit shown in figure 3.18 to reduce noise current through Z_c

The noise on V_{ss} may be reduced by choosing a low noise, low output impedance supply and maintaining low supply-impedance at high frequencies by the use of capacitors across the supply lines.

The impedance between BC can be kept low by avoiding connectors in this path and by the use of a ground plane.

3.3.3 Power supplies

As noted in the previous section, power supplies are notorious noise generators. In addition to the steady supply voltage there will be random noise arising from the rectifier and regulator circuits, power-line frequency harmonics, inadequately filtered and suppressed power-line borne noise (see next section), and modulation of the output resulting from the varying current demands of the supplied circuits. A typical power supply output is shown in figure 3.22.

This noise can be coupled to low signal level circuits through a number of routes. One route is that of common signal path as described in the previous section. Another is via a bias circuit as shown in figure 3.23. This type of noise may be reduced by attenuating it before it reaches the susceptible circuit – often by a simple CR low-pass filter. The addition of the capacitor shown will achieve this.

The use of a low noise power supply will clearly be advantageous. This may require the use of separate power supplies for low signal level circuits (figure 3.24). Such circuits usually require low supply current and low power supplies are generally less noisy than high power ones. In addition, this avoids the supply rail noise arising from circuits exhibiting high and rapid changes in current demand – for example, digital and switching circuits.

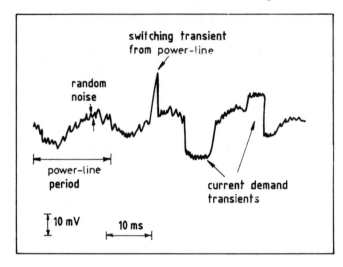

Figure 3.22   Typical noise superimposed on power supply output

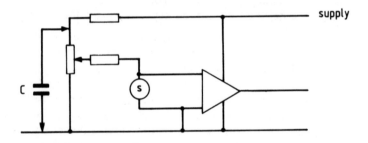

Figure 3.23   Coupling of supply noise to amplifier input via bias supply, and decoupling using low reactance capacitor C

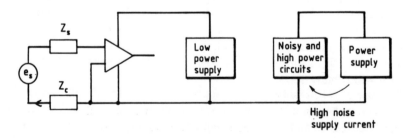

Figure 3.24   The use of separate power supplies to reduce conducted noise in low-level circuits

3.3.4 Power-line borne noise

The line supply has, superimposed on it, high frequency interference arising from devices coupled to the power-line and also large voltage transients from switching mechanisms.

The transients may be reduced in amplitude by the use of semiconductor transient suppressors – which are essentially back-to-back diodes with avalanche breakdown voltages, just above the peak line voltage (figure 3.25(a)) – or, alternatively, voltage-dependent resistors (varistors), which exhibit a rapidly decreasing AC resistance above a certain voltage drop (Standler, 1989). They provide low impedance pathways for the high voltage spikes. The devices obviously must be capable of passing very large currents of thousands of amps for a short period. The transient suppressor effectively acts together with the impedance of the power-line as a voltage divider. Its operation is aided by the fact that the impedance of the power-line at the high frequencies comprising the short duration interference spikes is much higher than it is at normal supply frequency (50–60 Hz) (Malack and Engstrom, 1976).

Radio frequency interference can be removed using a low-pass filter, as shown, which provides a low impedance pathway for the power-line frequency and a high series and low shunting impedance for the radio frequency interference. Note that low impedance paths via capacitors can be provided to earth on both live and neutral leads to remove interference which is common to both these leads (figure 3.25(b)).

High frequency interference can be capacitatively coupled between primary and secondary windings in a transformer. This interference can be reduced by providing an earthed metal foil screen between the windings. Note that the two ends of the foil must be insulated from one another to avoid creating a shorted turn in the transformer. This screening can be improved still further by having two metal foil screens insulated from one another, the screen nearest to the primary being taken to the power-line ground and the screen closest to the secondary being taken to the instrument 'ground' (figure 3.25(c)).

3.4 Noise of non-electrical origin

3.4.1 Introduction

The previous sections, 3.2 and 3.3, were concerned with electrical noise of electrical origin. Electrical signals from other circuits became 'noise' in the circuit of interest by means of unintended signal paths. In this section, noise which has a non-electrical origin is described. In most cases this noise is of mechanical origin – mechanical vibration of circuit boards, cables and

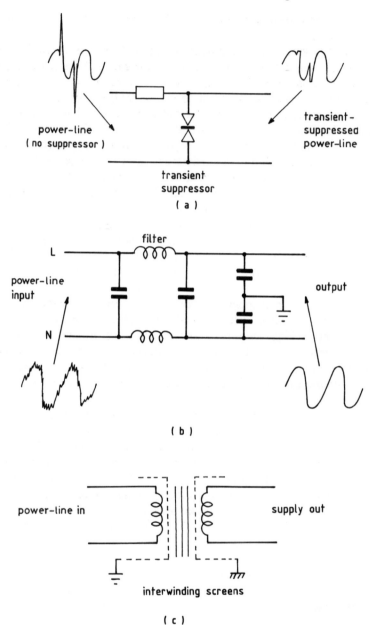

Figure 3.25 Methods of power-line borne interference reduction: (a) transient suppression; (b) radio frequency interference low-pass filter; (c) reduction of capacitative coupling between windings on power-line transformer

contacts, leading to electrical noise through mechano-electrical transduction of various types.

Movement or vibration of electrical circuits is often hard to avoid without special precautions. Instruments in vehicles, aircraft and rockets and some industrial plants can experience high levels of vibration. In addition, noise can be generated by movements of cables, by hand or automatically – for example, by the movement of a sensor or transducer coupled, by cable, to a stationary instrument.

The types of mechanically induced electrical noise, the susceptible circuit components and the choice of components and construction techniques used to minimise such noise are discussed in this section.

Also covered is the electrical noise resulting from temperature variation and from electrolytic contamination.

3.4.2 Triboelectric noise

The triboelectric effect is the transfer of charge when two materials are rubbed together and separated, leaving one positively and the other negatively charged. The phenomena of transfer of charge when two dissimilar insulators are rubbed together and the induction of charge density gradients in conductors by nearby charged insulators are well known. There is a transfer charge when any two materials come into contact, whether they are insulators or conductors, but in the case of two conductors the high electron mobility enables charge to equilibrate (electrons moving back to the electron-deficient, and therefore positively, charged material) as the conductors separate.

There is transfer of charge when insulators and conductors touch and separate, however, and this phenomenon can lead to noise in low signal level, high impedance circuits (Keithley *et al.*, 1984)

This is particularly noticeable in coaxial cables connecting very high impedance sources to very high impedance amplifiers (for example, electrometers). Flexing of such a cable leads to rubbing and intermittent contact between the conductors and the insulator leading to charge transfer, charge induction on the inner conductor and charging of the capacitor formed between the inner and outer conductors. As the resistance connected between the conductors (at either end of the cable) is reduced, the discharge rate of this capacitor is increased thereby reducing the charge developed and the noise voltage.

In order to illustrate the impedance levels at which this type of noise is a problem, the noise voltage across one metre of coaxial cable as it is flexed by hand is shown in figure 3.26 for two values of the resistance connected across the cable.

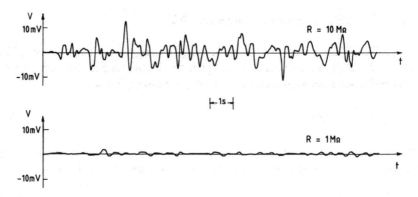

Figure 3.26 Noise emf across cable shunted by resistance *R* and flexed by hand

The frequency response of systems in which triboelectric noise is a problem is usually limited to very low frequencies as a result of the long time constant of the cable capacitance and the high resistance terminating the cable and/or by deliberate bandwidth limiting to reduce thermal noise.

The problem can be alleviated by reducing or eliminating cable flexing, but where such measures are not possible or insufficient to reduce the noise to acceptable levels, special low noise cable may be used. In one such cable there is a layer of graphite on the inner insulator between the insulator and the outer conductor as shown in figure 3.27. This reduces the friction between insulator and outer conductor, thereby reducing the incidence of local friction and separation which leads to charge transfer and also, by providing the insulator with a conducting surface, increasing the electron mobility on this surface and, thereby, the rate of charge equilibration – preventing the build-up of charge. It should be noted, however, that the use of such cable reduces but does not eliminate triboelectric noise – the cable is 'low' noise, not 'zero' noise.

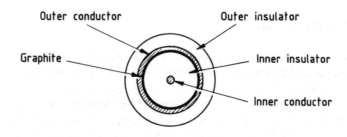

Figure 3.27 Cross-section of low noise cable

3.4.3 Conductor movement in magnetic field

An emf is induced in a conductor moving in a magnetic field (figure 3.28). The relationship between emf v_m, magnetic induction $\bar{B}$ and the velocity $\bar{v}$ of the conductor is given by equation (3.44) – Faraday's law:

$$v_m = \int (\bar{v} \times \bar{B}) d\bar{r} \qquad (3.44)$$

If $\bar{B}$ is taken as an average over a length l of straight conductor, the conductor is perpendicular to $\bar{B}$, and $\bar{v}$ makes angle θ with $\bar{B}$, then:

$$v_m = lBv \sin \theta \qquad (3.45)$$

Note that the magnetic field need not be static – movement of a conductor in the alternating field from a transformer will give rise to a movement-induced emf in addition to the voltage arising through mutual induction.

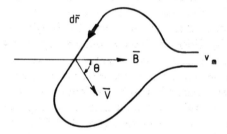

Figure 3.28 Noise emf generated by moving conductor in magnetic field. Note that, in general, $\bar{B}$ and $\bar{v}$ are functions of position along the conductor

Even the earth's weak magnetic field can lead to significant noise in moving cables. The emf generated in 1 m of cable moving at 0.01 m/s at right angles to a typical field of 4×10^{-5} Wb/m^2 is 0.4 µV.

3.4.4 Piezoelectric effect

The stressing of certain materials leads to a potential difference across electrodes attached to the surface of the material. This, and the inverse effect – the deformation of the material when a voltage is applied across the electrodes, is known as the piezoelectric effect. Materials with a large piezoelectric effect are used as electromechanical transducers. There are some commonly used insulating materials which have a small, but significant piezoelectric effect – ceramic insulators and some circuit board material, for example (Keithley *et al.*, 1984). Vibration of these materials can lead to noise voltages between attached conductors as illustrated in figure 3.29. The noise can be reduced if the mechanical vibration is reduced by the use of anti-vibration mounts, for example (figure 3.30), and by choosing insulating materials which exhibit a low piezoelectric effect.

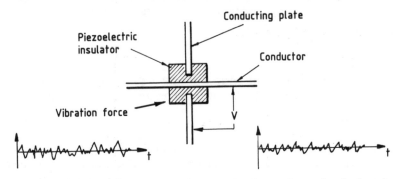

Figure 3.29 Electrical noise from transduction of mechanical noise in piezoelectric insulator

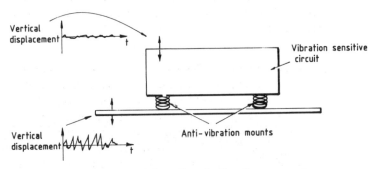

Figure 3.30   Use of anti-vibration mounts

3.4.5 Space charge effects

Any two conductors separated by an insulator form a capacitor. If the capacitor so formed contains charge Q, then the voltage (associated with this charge) across the conductors is:

$$V = Q/C \tag{3.46}$$

The capacitance C is a monotonically decreasing function of the conductor separation. The form of the relationship depends on the conductor geometry and the relative orientation. If the charge remains constant – no leakage – then a change in capacitance δC arising from a change in conductor separation leads to a change in capacitor voltage:

$$\delta V \simeq \frac{d(Q/C)}{dC} \, \delta C$$

$$= -\frac{Q}{C^2} \, \delta C \tag{3.47}$$

This is illustrated in figure 3.31.

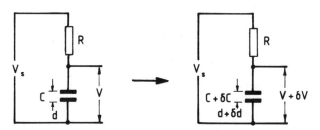

Figure 3.31　Change in potential difference across capacitor as a result of changing capacitor plate separation. If δd is positive, then δC is negative and δv is positive. Equilibration of capacitor potential difference to V_s takes place on timescale of the order of CR

A resistance R effectively in parallel with C will allow Q to change during the change in C, but only on a time scale of the order of CR. Changes in C on a time scale shorter than this will lead to the voltage change indicated in equation (3.47). If the time over which C changes is comparable or longer than CR, then the change in V will reduce as this time increases.

The effect described is the basis of the capacitor microphone. Sound pressure waves moving a microphone diaphragm forming one plate of a capacitor lead to capacitance changes and similar changes in the capacitor voltage. Likewise, vibration of conductors in close proximity, on a circuit board for example, and particularly in coaxial cables, where the capacitance is relatively high, can lead to sympathetic changes of the voltage between the conductors. This behaviour is known as microphony. It is illustrated in figure 3.32.

The problem can be alleviated by avoidance of mechanical vibration of critical circuit elements (including cable) and by reducing the steady voltage across cables carrying low-level signals.

3.4.6　Electrochemical

Noise emfs can result from electrochemical contamination of circuit boards. Weak batteries are formed from two dissimilar metals – which may be the

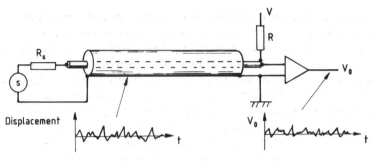

Figure 3.32   Cable microphony

copper in printed circuit board tracks and the lead in solder – and an electrolyte – which could be from solder flux on poorly cleaned boards.

The problem can be alleviated by thorough cleaning and covering the circuit boards with a moisture-resistant coating.

3.4.7   Thermal

Thermal emfs are developed when the junctions between two dissimilar metals are at different temperatures. In figure 3.33 the thermal emf e_{th} is proportional to the temperature difference $T_2 - T_1$. Variations in this temperature difference resulting from local variations in temperature, resulting in turn from turbulent convection (for example), lead to random variations in e_{th}. The metal junctions may be between copper PCB tracks and the gold plating on connectors, or between copper leads and tracks and lead/tin solder.

Another route by which temperature variation is converted to voltage variation is through the temperature dependence of resistance and semiconductor junction voltage drop.

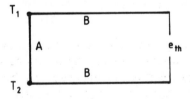

Figure 3.33   Thermal emf generated as a result of junctions between dissimilar metals A and B at different temperatures T_1 and T_2

The rate at which the temperatures of temperature-sensitive circuit elements change depends on the thermal capacity of the structures to which these elements are physically attached and the thermal conductivity to other structures. The variations are slow – in the sub-audio frequency range and the resultant change in circuit voltages is often termed 'drift'.

Thermo-electric emfs may be eliminated by avoiding junctions between dissimilar metals. This is not always easy, but reductions can be achieved by selecting metal junctions which exhibit low thermal emfs, using solder containing cadmium instead of lead, for example. The thermal emf at a copper–cadmium/tin junction is approximately $0.3\,\mu V/^{\circ}C$ compared with 1–$3\,\mu V/^{\circ}C$ at a copper–lead/tin junction (Keithley *et al.*, 1984).

Other precautions are the use, in susceptible circuits, of low temperature coefficient resistors, and amplifiers with balanced input circuits, such as that shown in figure 3.34, where the characteristics of the input transistors are closely matched and they are mounted close together, perhaps in the same package, to minimise temperature differences. Note that the matching is never perfect and amplifiers will always have a temperature-dependent input offset voltage and current, and that this is exacerbated by a lack of balance in the source circuit.

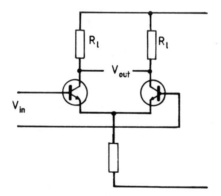

Figure 3.34 Use of balanced input circuit for an amplifier to reduce thermal noise or drift

All types of thermally induced noise are reduced by reducing temperature gradients and their variation, usually by mounting susceptible circuits on heat sinks with high thermal conductivity and heat capacity. The high thermal conductivity helps to minimise temperature gradients and high thermal capacity reduces the extent and rate of temperature variations (figure 3.35).

Only metals fulfil the requirements of high thermal conductivity and high thermal capacity in a reasonable volume. Insulators, generally, are poor heat conductors. Susceptible circuits must be thermally coupled to such heat sinks using thin insulators with thermal conductivities which are high for insulators. Suitable insulators are the anodising on aluminium, beryllium

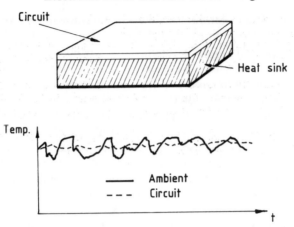

Figure 3.35 Use of heat sink to reduce circuit temperature fluctuations

oxide or epoxy resins with thermally conductive filler. The use of thermal insulation and shielding from convective air currents and high power, heat-emitting circuits is also advisable.

3.4.8 *Contact noise*

Randomly varying contact resistance in plugs, sockets, switches and poorly soldered joints can lead to noise by modulating supply, bias and signal currents. Such noise can be caused by a combination of mechanical vibration and poor contact and be affected by temperature variations leading to contraction and expansion at the contact. The noise may be avoided by making sure that contacts are clean and firmly in contact and that soldered joints are made carefully with adequate 'wetting' by the solder of each surface.

See also Chapter 4, section 4.4 on low frequency noise.

References

Baker, D., Koehler, D. C., Fleckenstein, W. O., Roden, C. E. and Sabia, R. (1970). *Physical Design of Electronic Systems, Vol. 1 Design Technology*, Prentice-Hall, Englewood Cliffs, New Jersey.

Corson, D. and Lorrain, P. (1962). *Introduction to Electromagnetic Fields and Waves*, Freeman, San Francisco.

Giacoletto, L. J. (1977). *Electronics Designers' Handbook*, McGraw-Hill, New York.

Hayt, W.H. (1981). *Engineering Electromagnetics*, 4th edn, McGraw-Hill, New York.

Keithley, J. F., Yeager, J. R. and Erdman, R.J. (1984). *Low Level Measurements*, 3rd edn, Keithley Instruments Inc, Cleveland, Ohio.

Malack, J. A. and Engstrom, J. R. (1976). 'RF impedance of United States and European power lines', *IEEE Transactions on Electromagnetic Compatibility*, **EMC-18**, 36–38.

Ruehli, A. E. (1972). 'Inductance calculations in a complex integrated circuit environment', *IBM Journal of Research and Development*, **16(5)**, 470–481.

Standler, R. B. (1989). *Protection of Electronic Circuits from Overvoltages*, Wiley, New York.

Terman, F. E. (1955). *Electronic and Radio Engineering*, McGraw-Hill, New York.

Weber, E. (1965). *Electromagnetic Theory*, Dover, New York.

4 Intrinsic Noise

4.1 Introduction

Noise from the sources described in the previous chapter may be reduced to negligible levels by attention to circuit layout and construction techniques. However, even if this noise were reduced to zero, some noise would remain. This remaining noise, intrinsic to the devices and even the conductors which make up the circuit, arises from the random motion of charge carriers. The fluctuating voltages and currents, being random, are only adequately described in terms of their statistics – their mean square values, probability density function and frequency spectra, for example.

As a result of its importance, and simply out of interest in such a fundamental phenomenon, there has been much theoretical and experimental investigation into intrinsic noise. However, it is beyond the scope of this book to describe this work in detail and readers who require more fundamental knowledge are referred to excellent reviews of the subject listed in the Bibliography. We describe here the source of intrinsic noise in sufficient detail to justify noise models of circuit elements and the description of methods of analysis of noise in circuits and systems.

4.2 Thermal noise

4.2.1 Introduction

Any conductor, even if it is not passing current as a result of connection to a signal source (including electromagnetic, magnetic or capacitive coupling) or power supply will have a fluctuating potential difference across its ends. This fluctuation or noise is caused by the random, thermally induced, motion of conduction electrons. This thermal noise is also known as Johnson or Nyquist noise after two of the early investigators of this phenomenon (Johnson, 1928; Nyquist, 1928).

The noise has a Gaussian pdf, as might be expected from a phenomenon resulting from a large number of random events, and has a spectral density:

$$S_t(f) = 4kTR \quad \text{V}^2/\text{Hz} \tag{4.1}$$

where k = Boltzmann's constant $(1.38 \times 10^{-23}\,\text{J/K})$
       R = the resistance of the conductor
and T = the temperature of the conductor on the absolute scale.

Since the spectral density is independent of frequency, the mean square voltage of the noise measured in bandwidth B is:

$$e_T^2 = \overline{e_t^2} = 4kTRB \quad \text{V}^2 \tag{4.2}$$

and the rms noise voltage is:

$$e_T = \sqrt{(\overline{e_t^2})} = (4kTRB)^{1/2} \quad \text{V} \tag{4.3}$$

The thermal emf (e_T) at a temperature of $17°\text{C}$ $(290\,\text{K})$ is shown as a function of resistance and bandwidth in figure 4.1.

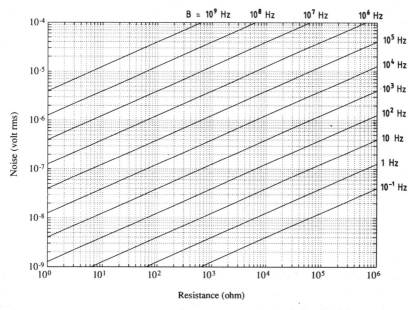

Figure 4.1   Thermal noise at 290 K versus resistance for different bandwidths

The resistance may be modelled, using the Thévenin model of a two-terminal network, as shown in figure 4.2(a). The real (noisy) resistance is equivalent to an ideal (noiseless) resistance in a series with a random voltage generator with a spectral density given by equation (4.1), or with an rms voltage (in a measurement bandwidth B) given by equation (4.3).

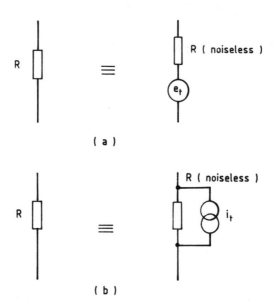

Figure 4.2   (a) Thévenin and (b) Norton equivalent circuits of a resistor

An equivalent (Norton) circuit with a random current generator in parallel with a noiseless resistance is shown in figure 4.2(b). In this case, the spectral density of the current generator is:

$$S_{ti}(f) = 4kT/R \quad \text{A}^2/\text{Hz} \tag{4.4}$$

or a current of:

$$i_T = \sqrt{\overline{i_t^2}} = (4kTB/R)^{1/2} \quad \text{A rms} \tag{4.5}$$

in a measurement bandwidth B.

4.2.2 Noise specification and dependence on bandwidth
(see appendix *C*)

Although the normalised power spectral density of the noise is as shown in equation (4.1) and measured in volt^2 per Hertz, it is often specified as its square root, measured in volts per square root Hertz. That is:

$$\mathcal{E}_t = S_t^{1/2}(f) = (4kTR)^{1/2} \quad \mathrm{V\,Hz^{-1/2}} \tag{4.6}$$

and the noise voltage measured in bandwidth B is:

$$e_T = \mathcal{E}_t B^{1/2} \quad \mathrm{V\ rms} \tag{4.7}$$

Similarly for current noise we can write:

$$\mathcal{I}_t = S_{ti}^{1/2}(f) \quad \mathrm{A\,Hz^{-1/2}} \tag{4.8}$$

and the noise current in bandwidth B is:

$$i_T = \mathcal{I}_t B^{1/2} \quad \mathrm{A\ rms} \tag{4.9}$$

4.2.3 Temperature dependence

The dependence of thermal noise on temperature means that the noise level may be reduced by lowering the temperature. Since it is dependent on the absolute temperature, however, a large reduction of temperature on the Centigrade scale is required to obtain a significant reduction. A variation in temperature of a few tens of degrees around room temperature is generally not significant. Writing the temperature on the Centigrade scale as T_c, the thermal emf (e_{TC}) at T_c compared with that at 17°C (e_{T17}) is, from equation (4.3):

$$
\begin{aligned}
e_{TC}/e_{T17} &= \left(\frac{4kRB(273 + T_c)}{4kRB(273 + 17)}\right)^{1/2} \\
&= \left(1 + \frac{T_c - 17}{290}\right)^{1/2} \\
&\simeq 1 + 0.0017(T_c - 17)
\end{aligned}
\tag{4.10}
$$

for small $T_c - 17$.

To obtain a significant reduction in noise it is necessary to cool to liquid nitrogen (77 K) or even liquid helium (4.2 K) temperatures. For example, a resistance of 1 kΩ has the noise emf in a bandwidth of 5 MHz, as shown in table 4.1

Table 4.1 Thermal noise across 1 kΩ in 5 MHz bandwidth

	Temperature (K)	*Noise emf* (μV)
Room	290	8.9
Liquid nitrogen	77	4.6
Liquid helium	4.2	1.1

4.2.4 Available power

It will be seen later, when the noise performance of systems involving power-matched devices is considered, that an important quantity is the available noise power from a noise source. This is the power dissipated in a matched load and can be easily calculated, by considering a load resistance R connected to the noise generator shown in figure 4.2(a), as

$$P_t = e_T^2/(4R)$$

$$= kTB \tag{4.11}$$

Note that it is not possible actually to obtain power from a thermal noise generator. Any connected resistive load generates its own thermal noise and there is no net power transfer.

4.2.5 Normalised power spectrum and noise bandwidth

It follows from equation (4.2) that, as the measurement bandwidth increases, the noise emf increases – apparently without limit. In fact, the spectral density formula in equation (4.1) is an approximation derived from classical thermodynamics. At very high frequencies, quantum effects should be taken into account and the product kT in equation (4.1) replaced by:

$$(kT)' = \frac{hf}{kT}(\exp(hf/(kT)) - 1)^{-1} \quad \text{J} \tag{4.12}$$

where h is Planck's constant (6.62×10^{-34} joule seconds). This factor leads to a progressive reduction in the spectral density at frequencies above kT/h. This frequency is O [10^{13} Hz] at room temperature. Since we are normally concerned with frequencies very much lower than this, the approximation $kT = (kT)'$ is very close.

In practice, the gain of the circuit in which the noise is measured reduces at high frequencies, either as a result of deliberate bandwidth limiting or as a result of stray reactive components (lead inductance, and stray capacitance). For example, consider the case of bandwidth limitation by a capacitance C in parallel with resistance R. The equivalent noise circuit is shown in figure 4.3.

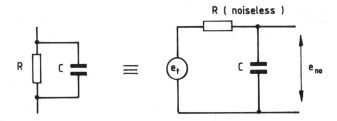

Figure 4.3 Equivalent circuit of resistor and capacitor in parallel

The transfer function of the filter formed by R and C is:

$$H(f) = \frac{\dfrac{-j}{2\pi f C}}{R - \dfrac{j}{2\pi f C}}$$

and the spectral density of e_{no} is (see chapter 2, section 2.4.4):

$$S_{no}(f) = |H(f)|^2 S_t(f)$$

$$= (1 + (2\pi f C R)^2)^{-1} 4kTR \tag{4.13}$$

This is shown, plotted for a fixed C and several values of R, in figure 4.4.

The total normalised power e_{No}^2 is found by integrating equation (4.13):

$$e_{No}^2 = \int_0^\infty (1 + (2\pi f C R)^2)^{-1} 4kTR \, df$$

$$= kT/C \quad \text{V}^2 \tag{4.14}$$

Thus e_{No} is independent of R and depends only on the capacitance C. For a fixed C as shown in figure 4.4, the bandwidth changes but the noise emf

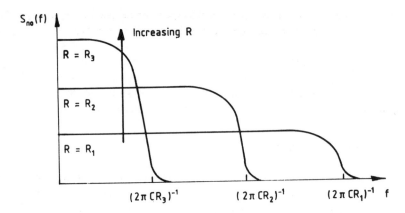

Figure 4.4 Spectrum of noise from resistor (R) and capacitor (C) in parallel

stays the same. Using equation (4.14) for the example of a capacitance of 1 pF at 290 K we find that e_{No} is 63 µV. If, however, we use the conventional 3 dB bandwidth for the CR low-pass filter in figure 4.3 $(B = (2\pi CR)^{-1})$ and use equation (4.3) to calculate the noise voltage, we obtain an erroneous figure of 50 µV.

The problem arises from a definition of bandwidth. The total noise in bandwidth B (equation (4.2)) is obtained by integrating the spectral density (a constant) over frequency range f_1 to f_2 such that $f_2 - f_1 = B$. This is equivalent to calculating the noise at the output of an ideal rectangular response filter.

In practice, the frequency response of the circuit determining the spectrum of the measured noise will not have sharp cut-off frequencies, and to obtain an accurate noise level we should integrate the spectral density of the noise at the circuit output. With reference to figure 4.5(a), the mean square output noise voltage is:

$$e_{\text{No}}^2 = \int_0^\infty G_p(f) S_t(f)\, df$$

$$= 4kTR \int_0^\infty G_p(f)\, df \qquad (4.15)$$

where:

$$G_p(f) = |H(f)|^2 \qquad (4.16)$$

is the normalised power gain of the circuit.

If G_{po} is the mid-band gain for a bandpass filter or low frequency gain for a low-pass filter, then we can define a noise bandwidth (Haus *et al.*, 1960):

$$B_n = \frac{\int_0^\infty G_p(f)df}{G_{po}} \tag{4.17}$$

in which case:

$$e_{No} = (4kTRB_nG_{po})^{1/2} \tag{4.18}$$

Comparing this result with equation (4.3) we see that this is the noise level expected at the output of a rectangular filter with a passband gain of G_{po} and bandwidth B_n. The noise bandwidths of low-pass and bandpass filters are illustrated in figures 4.5(b) and 4.5(c) respectively. An alternative view, following from equation (4.17), is that we are replacing the actual filter by a rectangular filter having the same low frequency (low-pass filter) or mid-band (bandpass filter) gain and the same area under the frequency response curve.

In order to obtain accurate estimates of noise levels, therefore, the bandwidth used in equation (4.3) should be calculated from the frequency response of the system between the noise source and the measurement point. If we apply this to the case of the low-pass filter defined by a single CR network (figure 4.3), then:

$$e_{No} = (4kTRB_n)^{1/2}$$

$$G_{po} = 1$$

$$G_p(f) = (1 + (2\pi fCR)^2)^{-1} \tag{4.19}$$

and equation (4.17) gives a noise bandwidth:

$$B_n = (4CR)^{-1} \tag{4.20}$$

Since the 3 dB bandwidth of such a filter is $B_3 = (2\pi CR)^{-1}$, then:

$$B_n = \frac{\pi}{2}B_3 \tag{4.21}$$

In the case considered, if we multiply the 3 dB bandwidth by $\pi/2$ before calculating the noise voltage we obtain $63\,\mu V$, which agrees with the result obtained using equation (4.14).

In these examples there is no difficulty over the frequency at which G_{po} is measured. In many cases, however, the frequency response of a bandpass

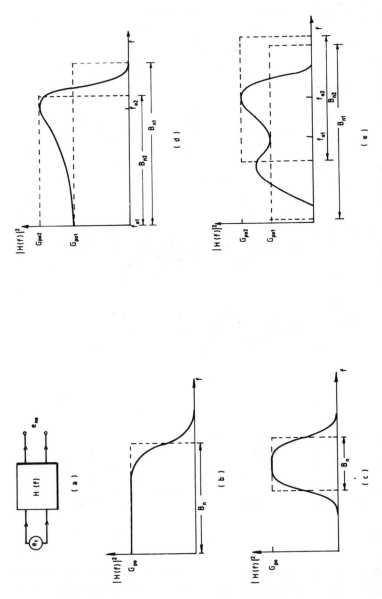

Figure 4.5 Noise bandwidth of thermal noise filtered by circuit with frequency response $H(f)$: (a) circuit; (b) low-pass filter; (c) bandpass filter; (d) low-pass filter with high frequency peak; (e) bandpass filter, asymmetric and double-humped response

filter is not symmetrical and the response of a low-pass filter is not maximum at zero frequency. Examples of these cases are shown in figures 4.5(d) and 4.5(e). This leads to some uncertainty over the measurement of noise bandwidth. The bandwidths B_{n_1} and B_{n_2}, calculated for two possible choices, f_{o_1} and f_{o_2}, of frequencies at which the gain G_{po} can be measured, are shown. In practice, the decision of which bandwidth to adopt is arbitrary and does not affect the final noise calculation. The product $G_{po}B_n$ is the same whatever frequency is chosen for the measurement of G_{po} and it is this product which determines the calculated noise voltage (equation (4.18)).

4.2.6 Thermal noise from mixed resistive and reactive sources

It has been determined that the noise from any impedance is determined by the resistive component or real part of that impedance (Nyquist, 1928). This may be extended to any passive network of resistive and reactive components. The noise emf measured across any two terminals in the network is determined by the real part of the impedance measured across those terminals. This is illustrated in figure 4.6.

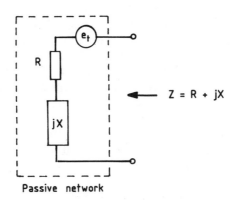

Passive network

Figure 4.6   Thermal noise from passive network

Note that, in general, Z, X and R are functions of frequency. This means that the spectral density (equation (4.1)) is no longer constant and the rms noise voltage in the frequency range f_1 to $f_1 + B$ is:

$$e_T = \left(4kT \int_{f_1}^{f_1+B} R(f)\, df \right)^{1/2} \tag{4.22}$$

or the output of a system with power gain $G_p(f)$:

$$e_T = \left(4kT \int_0^\infty G_p(f)R(f)\,df\right)^{1/2} \tag{4.23}$$

These reduce to equations (4.3) and (4.18) respectively if R is independent of frequency.

Note that if the resistive components in the network are at different temperatures, the temperature to be used in the above formulae is a weighted mean of the temperatures of the individual resistances. This will be discussed in more detail in the next chapter.

4.2.7 *Non-electrical sources and equivalent circuits*

Although this book is concerned with electrical noise, it is worth noting that thermal noise arises in any energy-dissipative device. For example, the absolute limit of sensitivity of microphones is determined by the noise caused by random bombardment of the microphone diaphragm by air molecules, and this can be related to the resistance to motion caused by the same bombardment (Becking and Rademakers, 1954).

The thermal noise at the electrical output of the microphone may be calculated, as above, using the real part of the impedance presented at the electrical output. The impedance is calculated from the equivalent circuit of the microphone which, in addition to any electrical resistances (such as series resistance of the coil in a moving coil microphone), includes the electrical resistance representing the resistance to motion of the diaphragm.

The noise at the electrical terminals of any device containing non-electrical components can be calculated from an equivalent circuit and it is clearly important that the electrical resistances representing dissipative non-electrical components are included if they are significant. For example, resistance should be included to represent the effect of dielectric loss in capacitors, and hysteresis and eddy current loss in the magnetic cores of inductors and transformers.

4.3 Shot noise

In addition to the thermal noise described in the previous section there are various types of noise which result from the passage of current through circuit elements. One of these – shot noise – results from the random passage of individual charge carriers across a potential barrier. It was first noted in thermionic valves where the potential barrier concerned is at the valve cathode (Schottky, 1918). The current flow across a semiconductor junction is more complex, but the current flowing across the junction exhibits the same level of noise (van der Ziel and Chenette, 1978; Buckingham, 1983).

The noise current has a constant spectral density of:

$$S_i(f) = 2eI_{DC} \quad A^2/Hz \tag{4.24}$$

where e is the electronic charge $(1.60 \times 10^{-19}\,C)$ and I_{DC} is the average current. The noise current is a small random fluctuation of the current flowing. If $i(t)$ is the total current then:

$$i(t) = I_{DC} + i_n(t) \tag{4.25}$$

where the spectral density of the noise current $i_n(t)$ is given by equation (4.24). This is illustrated in figure 4.7.

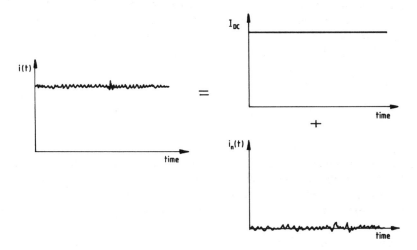

Figure 4.7   Current through semiconductor junction as sum of steady and noise currents

Since shot noise arises from the sum of a large number of individual events, the noise has a Gaussian pdf. As the spectral density is constant, all the comments and calculations related to thermal noise spectra and bandwidth also apply to shot noise. In particular, the rms noise current measured in noise bandwidth B_n is:

$$i_N = (2eI_{DC}B_n)^{1/2} \quad A \text{ rms} \tag{4.26}$$

The noise current for range of direct currents and bandwidths is shown in figure 4.8.

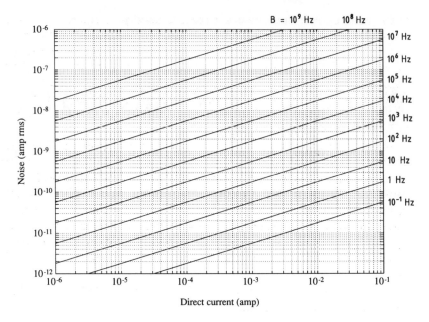

Figure 4.8   Shot noise current versus direct current and noise bandwidth

4.4   Low frequency (excess) noise

4.4.1   Introduction

When current passes through a resistor or a semiconductor, noise is generated in excess of the thermal noise in a resistor or the thermal noise plus shot noise in a semiconductor. This 'excess' noise has a spectral density which increases as the frequency decreases and, because it is most noticeable at low frequency, it is also known as 'low frequency' noise. In addition, it is known as 'flicker' noise because of the flickering of the needle on a meter measuring current with a significant component of this type of noise, or '$1/f$' (one-over-eff) noise since its power spectrum has an approximate $1/f$ dependency, and used to be known as 'semiconductor' noise since the noise in early semiconductor devices was dominated by this type.

The mechanism giving rise to this noise is still not clearly understood and indeed it is possible that there is more than one mechanism leading to similar effects. What is clear is that in resistors the magnitude of the noise is strongly related to the number and quality of the contacts within the resistor. The excess noise is highest in carbon composition resistors which consist of compact carbon granules, and where the current has to pass through many points of contact between individual granules. This type of resistor is generally not used in low noise design. Metal film resistors have lower excess noise and wire-wound the lowest of all. In these resistors, the excess noise is determined by the quality of the end connections. There is experimental evidence that this type of noise in resistors results from a random variation in resistance (Voss and Clarke, 1976; Beck and Spruit, 1978). This variation manifests itself as a variation in voltage drop across a resistance when current is flowing.

In semiconductors, low frequency excess noise can be generated as a result of fluctuations in the number of charge carriers – resulting, in turn, from random movements in and out of carrier traps associated with crystal imperfections, particularly at the crystal surface. Low frequency noise in semiconductors has reduced over the years as manufacturing techniques have improved. As with shot noise, low frequency noise in semiconductors appears as a fluctuation of the steady current flowing through the device.

With both resistors and semiconductors there is a wide variation in low frequency noise both between types and between samples of a particular type. Only maximum or average levels for a particular type are given by the manufacturer.

4.4.2 Noise characteristics

Excess noise has a Gaussian pdf and a spectral density:

$$S_f(f) = K_f/f^\alpha \quad \text{V}^2/\text{Hz} \tag{4.27}$$

where K_f is dependent on the current passing through the device and is equal to the spectral density at 1 Hz. The form of the spectrum is shown in figure 4.9. The index α is usually in the range 0.8–1.4 (Wolf, 1978) and for approximate calculation purposes is often taken as unity. In this case:

$$S_f(f) = K_f/f \quad \text{V}^2/\text{Hz} \tag{4.28}$$

The mean square voltage of the noise in the frequency range between f_1 and f_2 is:

$$e_F^2 = \int_{f_1}^{f_2} (K_f/f) \, df$$

$$= K_f \ln(f_2/f_1) \quad \text{V}^2 \tag{4.29}$$

Unlike thermal and shot noise where the noise power is proportional to bandwidth, $1/f$ noise power is dependent on the ratio of the upper and lower

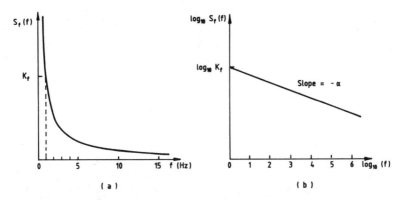

Figure 4.9 $1/f$ noise spectrum: (a) lin–lin plot; (b) log–log plot

frequency limits. The noise power in each frequency octave $(f_2/f_1 = 2)$ or decade $(f_2/f_1 = 10)$ is constant. The noise in a fixed bandwidth decreases as the centre frequency increases. This means that this type of noise is significant only at low frequencies and, above some frequency, dependent on the value of K_f and the thermal and/or shot noise, $1/f$ noise is negligible compared with the constant density noise.

Typical $1/f$ noise waveforms compared with those of thermal noise in three frequency decades are shown in figure 4.10.

The frequency decade is commonly used when dealing with $1/f$ noise. The normalised noise power in each decade (from equation (4.29)) is:

$$e_{Fd}^2 = K_f \ln(10)$$

$$= 2.30K_f \quad V^2(\text{frequency decade})^{-1} \tag{4.30}$$

or:

$$e_{Fd} = 1.52K_f^{1/2} \quad V(\text{frequency decade})^{-1/2} \tag{4.31}$$

Note that equations (4.28)–(4.31) are valid only for $\alpha = 1$; for more accurate calculations of noise power it is necessary to integrate $S_f(f)$ using equation (4.27). In many cases it will be sufficiently accurate to use the 3 dB cut-off frequencies for f_2 and f_1 in equation (4.29), but again, for more accurate calculation, the frequency response of the circuit under investigation should be used. The mean square noise voltage resulting from a $1/f$ noise source when the normalised power gain between noise source and measurement point is $G_p(f)$ is given by:

$$e_F^2 = \int_0^\infty S_f(f)G_p(f) \, df \tag{4.32}$$

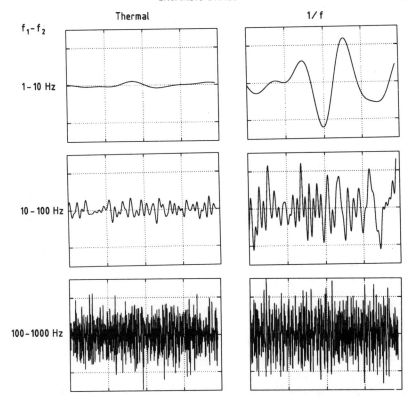

Figure 4.10   Thermal and $1/f$ noise signals in three frequency decades

Since this noise will be present at the same time as thermal (and maybe shot) noise, the total noise will be:

$$e_N^2 = \int_0^\infty (S_f(f) + S_t(f))G_p(f)\, df \tag{4.33}$$

Note that, since the spectral density is not constant, the equivalent noise bandwidth cannot be used as in the case of thermal and shot noise.

There is clearly a potential difficulty in the form of the spectrum in that it tends to infinity as the frequency tends to zero. This difficulty led to the supposition that there might be some low frequency limit to this increase in spectral density. However, experimental investigations at progressively lower frequencies have not uncovered such a limit. It might be thought that this leads to an infinite $1/f$ noise power in circuits with a passband that extends down to zero frequency (DC). There is, however, a high-pass filtration intrinsic in any circuit which is determined by the length of time that the circuit is switched on. The circuit cannot respond to frequency components with periods much greater than this time duration.

4.4.3 Excess noise in resistors

It is found that the excess noise in resistors is proportional to the direct current flowing and resistance (that is, to the DC voltage drop), therefore we can write:

$$e_{Fd} = C_F V_{DC} \quad \text{V(frequency decade)}^{-1/2} \tag{4.34}$$

where C_F = rms noise voltage (DC voltage drop)$^{-1}$ (frequency decade)$^{-1/2}$.

This is how manufacturers usually specify the excess noise voltage for resistors. C_F is typically in the range 0.01–1 μV V_{DC}^{-1} (frequency decade)$^{-1/2}$. Alternatively, it can be expressed in dB as a noise index (Conrad *et al.*, 1960):

$$NI = 20 \log_{10}\left(\frac{e_{Fd} \times 10^6}{V_{DC}}\right) \quad \text{dB in one frequency decade} \tag{4.35}$$

where (in this case) e_{Fd} is in microvolts.

Thus a resistor with an excess noise of 1 μV V_{DC}^{-1} (frequency decade)$^{-1/2}$ has a noise index of 0 dB.

Note that the frequency range is not necessarily an integer number of decades. In general, the number of decades is calculated from:

$$\text{No. of frequency decades} = \log_{10}(f_2/f_1) \tag{4.36}$$

As an example, the excess noise in a frequency range of 10 Hz to 30 kHz from a 10 kΩ resistor with an NI of zero dB and a voltage drop of 10 V is:

$$1\,\mu\text{V} \times 10 V_{DC} \times (\log_{10}(30 \times 10^3/10))^{1/2} = 18.6\,\mu\text{V}$$

For comparison, the thermal noise is 2.2 μV.

4.5 Burst (popcorn) noise

Burst or popcorn noise is a sudden and short-duration change in the current through a semiconductor junction (Jaeger and Brodersen, 1970; Hsu, 1971a, b; Koji, 1975). A typical burst current wave form is shown in figure 4.11(a). Usually the current switches between only two levels, as shown, but switching between more levels is not unknown. This noise is always mixed with thermal, shot and maybe $1/f$ noise, which may mask it. The appearance when they are of comparable levels is shown in figure 4.11(b). The difference between the two current levels is typically 10^{-8} amp and the duration of each burst is from a few microseconds upwards. The spectrum of this noise is similar to the low frequency noise discussed in the previous section, being proportional $1/f^\alpha$ where α is in the range 1–2, and usually approximately 2. Unlike $1/f$ noise the spectrum does not increase indefinitely as the frequency decreases but levels off at a frequency dependent on the frequency of transition between the two current levels.

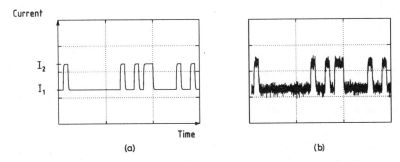

Figure 4.11 (a) Burst noise current and (b) burst noise with thermal noise

The pdf of this noise is clearly not Gaussian and is shown diagrammatically in figure 4.12. The relative heights of the two peaks are determined by the average dwell times at each level. Note that, unlike Gaussian noise, the pdf shape is dependent on the circuit bandwidth, since this determines the time spent in transition between levels.

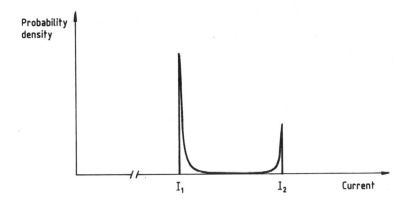

Figure 4.12   Probability density function of burst noise

Popcorn noise is associated with crystal defects and one explanation is that it is a modulation of the current through a defect by the random occupancy of charge carrier traps associated with the defect. It is usually only found in a small fraction of samples of any device and it may be avoided by selection.

References

Beck, H. G. E. and Spruit, W.P. (1978). '1/f noise in the variance of Johnson noise', *Journal of Applied Physics*, **49(6)**, 3384–3385.

Becking, A. G.T. and Rademakers, A. (1954). 'Noise in condenser microphones', *Acustica*, **4(1)**, 96–98.

Buckingham, M. J. (1983). *Noise in Electronic Devices and Systems*, Ellis Horwood, Chichester.

Conrad, G. T., Newman, N. and Stansbury, A. P. (1960). 'A recommended standard resistor-noise test system', *IRE Transactions on Component Parts*, **CP-7**, 71–88.

Haus, H. A. *et al.* (1960). 'IRE standards on methods of measuring noise in linear twoports, 1959', *Proceedings of the IRE*, **48**, 60–68.

Hsu, S.T. (1971a). 'Noise in high gain transistors and its application to the measurement of certain transistor parameters', *IEEE Transactions on Electron Devices*, **ED-18(7)**, 425–431.

Hsu, S.T. (1971b). 'Bistable noise in operational amplifiers', *IEEE Journal of Solid-State Circuits*, **SC-6(6)**, 399–403.

Jaeger, R. C. and Brodersen, A. J. (1970). 'Low frequency noise sources in bipolar junction transistors', *IEEE Transactions on Electron Devices*, **ED-17(2)**, 128–134.

Johnson, J. B. (1928). 'Thermal agitation of electricity in conductors', *Physics Review*, **32**, 97–109.

Koji, T. (1975). 'The effect of emitter-current density on popcorn noise in transistors', *IEEE Transactions on Electron Devices*, **ED-22**, 24–25.

Nyquist, H. (1928). 'Thermal agitation of electric charge in conductors', *Physics Review*, **32**, 110–113.

Schottky, W. (1918). 'Über spontane Stromschwankungen in verschiedenen Elekrizitätsleitern', *Ann. Phys. (Leipzig)*, **57**, 541–567.

van der Ziel, A. and Chenette, E. R. (1978). 'Noise in solid state devices', *Advances in Electronics and Electron Physics*, **46**, 313–383.

Voss, R. F. and Clarke, J. (1976). 'Flicker (1/f) noise: equilibrium temperature and resistance fluctuations', *Physical Review B*, **13(2)**, 556–573.

Wolf, D. (1978). '1/f-noise: Noise in physical systems', in *Proceedings of the 5th International Conference on Noise, Bad Nauheim, 1978* (Wolf, D., ed.), Springer-Verlag, Berlin, pp. 122–133.

5   Noise Circuit Analysis

5.1   Introduction

In any circuit there are a number of noise sources operating simultaneously. There is thermal noise from resistors and from the resistive elements of other components, shot noise from semiconductor devices, excess $(1/f)$ noise from both the above, and perhaps popcorn noise from the semiconductors. In order to deal with the problem of noise both from a design and a trouble-shooting point of view, we need a method of analysis and some standard method for comparing performance. In particular, we need to be able to calculate the net effect of a (possibly) large number of noise sources within a circuit. It would be useful in this exercise to have noise models of components or even whole circuit modules such as amplifiers, and in order to compare the performance of different circuits we need recognised figures of merit. In addition, we should like to be able to use these analysis techniques in order to optimise performance.

The noise descriptors used and their relationships are shown in appendix C.

Restricting the discussion to linear networks means that the standard techniques of analysis of these networks may be used in the analysis of noise. In particular, in order to find the net effect of a large number of noise sources we make use of the principle of superposition, which states that in a linear network the response for two or more sources acting simultaneously is the sum of the responses for each source acting alone, with the other voltage sources short-circuited and current sources open-circuited.

The noise at the output port of a network may, therefore, be calculated by adding the outputs arising from each source acting in isolation. Usually, the individual noise sources are independent, and therefore uncorrelated, and the resultant signal spectral density is found by adding the spectral densities of the individual outputs. For example, if $S_{oi}(f)$ is the normalised spectrum at the output arising from each source $i = 1, 2 \ldots n$, then the net output spectrum is:

$$S_{on}(f) = \sum_{i=1}^{n} S_{oi}(f) \tag{5.1}$$

where:

$$S_{oi}(f) = S_i(f)G_{pi}(f) \qquad (5.2)$$

$S_i(f)$ is the normalised power spectrum of noise generator i and $G_{pi}(f)$ is the normalised power gain between the generator and the output.

Integrating equation (5.1) with respect to frequency gives the normalised output power:

$$e_{on}^2 = \sum_{i=1}^{n} e_{oi}^2 \qquad (5.3)$$

where e_{oi} is the rms output with generator i acting alone. Since the sum involves squared quantities it is easy for some sources to dominate the output such that noise sources giving rise to small outputs can be ignored. For example, if two sources A and B are involved, A gives rise to an output of e_{oa} and B an output of $3e_{oa}$, then neglecting the contribution from source A leads to only a 5 per cent error in the net rms output (see section 2.2.9).

Equations (5.1)–(5.3) are strictly valid only if there is no correlation between any two noise sources. This will be approximately true if the correlation of any pairs of noise sources is small and/or the contribution of those sources to the sum is small.

There are occasions when we need to deal with partially correlated noise sources – when analysing the noise model of high frequency amplifiers, for example. It will be shown in section 5.2.14 that we can deal with this problem by splitting one of the pair of correlated generators into a fully correlated and an uncorrelated generator. By combining the two correlated generators, we end up with two generators which are uncorrelated. The process may be repeated for other correlated pairs.

5.2 Networks

5.2.1 Introduction

It is often easier to deal with networks in terms of their properties as measured at their ports rather than from a knowledge of their internal structure. This is also true of the noise performance.

5.2.2 Two-terminal or one-port networks

Any one-port, passive, linear network of resistors, capacitors and inductors will present, at its port, an impedance Z which may be considered as a resistance R (real part of Z) and reactance X (imaginary part of Z), and a thermal noise emf. The model of the network is shown in figure 5.1(a). If the resistors in the network are all at the same temperature T – that is to say, they are in thermal equilibrium – then the noise spectral density of e_{on} is that given by a resistance R at temperature T (Nyquist, 1928):

$$S_{on}(f) = 4kTR(f) \quad \mathrm{V^2/Hz} \tag{5.4}$$

This may be established, using a thermodynamic argument which, stated briefly, is that if a resistance R at temperature T is connected across the port, then, since the resistors are in thermal equilibrium and there is no power dissipated in a reactive component, there is no net transfer of power and the thermal emf of the load must equal the thermal emf of the network. Note that, if the network contains reactive components, then in general, the resistance is a function of frequency and therefore the spectral density is not constant.

The equivalent circuit shown in figure 5.1(a) is the Thévenin equivalent circuit. In the Norton equivalent circuit, the admittance $(Y = 1/Z)$ at the port is the sum of a conductance G and susceptance B as shown in figure 5.1(b). The spectral density $S_{oni}(f)$ of the equivalent noise current generator i_{on} is determined by the conductance. Thus:

$$S_{oni}(f) = 4kTG(f) \quad \mathrm{A^2/Hz} \tag{5.5}$$

If the resistors in the network are at different temperatures, then the temperature to be used in equations (5.4) and (5.5) will be a weighted mean of the temperatures of the individual resistors and be determined by their temperatures and the contribution of each to the port resistance. Given the insensitivity of thermal emfs to variations of a few tens of degrees Centigrade around room temperature or higher, the calculation of this mean temperature is unlikely to be necessary in practice.

As an example, the single CR network in figure 4.3 can be analysed as above. The impedance across the terminals of the network is:

$$Z = (1/R + j2\pi fC)^{-1}$$

$$= R(1 + j2\pi fCR)^{-1}$$

$$= \frac{R(1 - j2\pi fCR)}{1 + (2\pi fCR)^2} \tag{5.6}$$

The real part of Z is:

$$\mathrm{Re}[Z] = R(1 + (2\pi fCR)^2)^{-1} \tag{5.7}$$

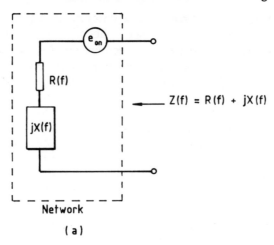

(a)

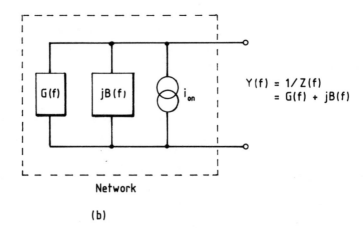

(b)

Figure 5.1   Noise equivalent circuits of one-port network: (a) Thévenin;
            (b) Norton

and the normalised spectral density of the noise voltage measured across the
terminals is (from equation (5.4)):

$$S_{on}(f) = 4kTR(1 + (2\pi fCR)^2)^{-1} \tag{5.8}$$

which is identical to that calculated treating the CR network as a low-pass
filter filtering the thermal noise from R (equation (4.13)).

5.2.3 Noise temperature

If the network is not passive but contains non-thermal noise sources (for example, shot and $1/f$ noise), then the noise measured at the port will be higher than that expected from a calculation of thermal noise. The noise level may be expressed as an equivalent noise temperature T_n. This is the temperature at which the resistive component of the port impedance gives thermal noise equal to the measured noise. If the measured noise spectral density is $S_{mv}(f)$ then the noise temperature is:

$$T_n(f) = \frac{S_{mv}(f)}{4kR(f)} \qquad (5.9)$$

or, if $R(f)$ is constant over bandwidth B, and e_M is the rms noise voltage measured in bandwidth B (from f_1 to $f_1 + B$):

$$T_n = \frac{e_M^2}{4kTRB} \qquad (5.10)$$

where:

$$e_M^2 = \int_{f_1}^{f_1+B} S_{mv}(f)df \qquad (5.11)$$

5.2.4 Spot frequency and broadband measurements

Note that equation (5.9) gives a value of T_n in terms of the noise spectral density at frequency f. In this case, T_n is a function of frequency and is known as the 'Spot frequency' noise temperature. Equation (5.10) gives a value of T_n which uses the normalised power e_M^2 measured in the frequency band f_1 to $f_1 + B$, and is known as the average or broadband noise temperature (Haus *et al.*, 1960a). There are a number of indices of noise performance of which spot frequency or broadband measurements may be made. In order to avoid multiplicity of subscripts the same variable name will be used for both, but shown as a function of frequency, or described as such, in the case of a spot frequency measurement.

5.2.5 Equivalent noise resistance

Another concept which is sometimes useful in simplifying equations is the equivalent noise resistance of a noise generator (Faulkner, 1968). This is the value of the resistance having a thermal noise equal to the noise of the generator at a standard temperature (usually 290 K). Thus the spot frequency noise resistance $R_{mv}(f)$ of the one-port network considered in the previous section is given by:

$$4kTR_{mv}(f) = S_{mv}(f)$$

or:

$$R_{mv}(f) = \frac{S_{mv}(f)}{4kT} \tag{5.12}$$

and the broadband equivalent noise resistance by:

$$R_{mv} = \frac{e_M^2}{4kTB} \tag{5.13}$$

The equivalent spot frequency and broadband noise resistances of a current generator of measured spectral density $S_{mi}(f)$ and rms current i_M in band B are (from equation (4.4)):

$$R_{mi}(f) = \frac{4kT}{S_{mi}(f)} \tag{5.14}$$

$$R_{mi} = \frac{4kTB}{i_M^2} \tag{5.15}$$

Note that equivalent noise resistances are simply convenient concepts and not real resistances. They cannot, of course, be used as substitutes for noise generators in the circuit to be analysed.

5.2.6 Two-port networks

Many of the circuit modules whose noise performances are of interest have inputs and outputs – they are two-port networks. By considering each port in turn as the connection to a one-port network with the other port left open-circuit, the real network can be seen as equivalent to an ideal (noiseless) network (with the same response to signal inputs) together with voltage noise generators in series with each port, as shown in figure 5.2(a) (Montgomery, 1952). A similar argument leads to an alternative equivalent circuit with parallel current noise generators at each port, as shown in figure 5.2(b). In this case, the noise current at each port is that measured or calculated with the other port shorted. It should be noted that, in general, the noise at each port arises from different contributions of the same internal noise generators, and the equivalent noise generators at the two

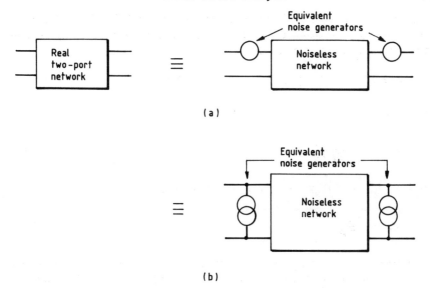

Figure 5.2   Equivalent circuits of a two-port network with noise generators at each port: (a) voltage generators; (b) current generators

ports are partially correlated. However, in many networks, particularly low frequency amplifiers the correlation is small and may be ignored.

It is shown in appendix B that an alternative equivalent circuit is that shown in figure 5.3, where equivalent generators – in this case, a voltage and a current generator – are both located at one of the ports – in this case, the input port (Becking *et al.*, 1955; Rothe and Dahlke, 1956).

This equivalent circuit has the advantage, when analysing amplifiers, that the source signal generator, source noise generator and equivalent amplifier noise generators are located at one place in the circuit. Because the response of the amplifier to signals and noise sources is the same, it is often possible to analyse noise performance without taking the amplifier gain into account. The circuit in figure 5.3 is a suitable noise model for an amplifier and we will refer to it as such, although the model and analysis applies to two-port networks in general.

It must be noted that this equivalent circuit is equivalent only so far as the noise at the output port is concerned. It cannot be used to calculate the actual noise in the input circuit. Since it is the relative levels of signal and noise at the output which are normally of interest, this is not a severe limitation in practice. The remarks on the correlation of the generators shown in figure 5.2 also applied to the two generators shown in figure 5.3.

It is possible to analyse the general case of the amplifier with two significantly correlated generators, treating the zero or insignificant

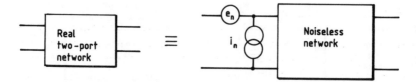

Figure 5.3   Equivalent circuit of a two-port network with two noise
              generators at one (input) port

correlation case as a special instance. However, it is probably easier to
follow if the simpler, and more widely used, uncorrelated generator
approximation is analysed first, followed by the more general case.

In order to aid comparison with other texts, the analysis (following
Faulkner, 1968; Letzter and Webster, 1970) is carried out without using the
spectral density of the noise source specifically, although certain key results
will be additionally expressed in terms of spectral densities.

The rms voltage of a noise generator measured in narrow bandwidth Δf
centred at frequency f will be denoted by e with an upper case first subscript
(for example, e_N) and similarly for a current generator (i_N). The bandwidth
Δf will always be sufficiently small so that the spectral density and circuit
frequency responses change negligibly over the bandwidth. This corresponds to the practice of 'Spot' frequency noise measurements. If $S_{nv}(f)$ is
the spectral density of the above noise generator then:

$$S_{nv}(f) = e_N^2 \Delta f^{-1} \quad \text{V}^2/\text{Hz} \tag{5.16}$$

The normalised spectral density of noise sources is frequently expressed as
its square root. Thus the root spectral densities of the equivalent input noise
generators shown in figure 5.3 are, for the series voltage generator:

$$\mathcal{E}_n = S_{nv}^{1/2}(f)$$

$$= e_N \Delta f^{-1/2} \quad \text{V Hz}^{-1/2} \tag{5.17}$$

and for the parallel current generator:

$$\mathcal{I}_n = S_{ni}^{1/2}(f)$$

$$= i_N \Delta f^{-1/2} \quad \text{A Hz}^{-1/2} \tag{5.18}$$

They represent the noise voltage and current respectively, measured in
narrow band Δf, centred on frequency f. They are usually written, as above,
without their frequency dependence being specifically indicated.

A broadband or average noise level will be indicated by the use of a bandwidth B. In order to avoid too many subscripts, the same variables e_N and i_N, for example, will be used to denote the rms noise voltage and current, respectively, measured in this wider frequency band. Whether a spot or broad band quantity is meant will be obvious from the context.

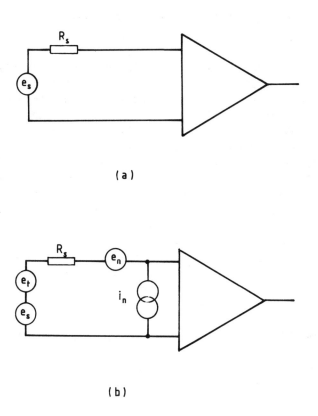

(a)

(b)

Figure 5.4 (a) Signal source e_s with source resistance R_s connected to amplifier and (b) equivalent noise circuit

Consider an amplifier connected to a signal source with source resistance R_s as shown in figure 5.4(a). Replacing the source resistance and the amplifier by their noise models gives the equivalent circuit shown in figure 5.4(b). The thermal noise in R_s is represented by voltage generator e_t. The circuit at the input (figure 5.5(a)) may be converted into the equivalent circuit shown in 5.5(b) by converting the current generator i_n to its equivalent voltage generator:

$$e_i = i_n R_s \tag{5.19}$$

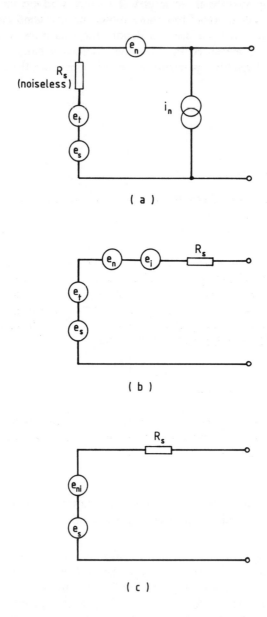

Figure 5.5   Noise equivalent amplifier input circuits

Combining the (uncorrelated) noise sources into one generator gives the circuit in figure 5.5(c) where, since the generators are uncorrelated (see section 2.2.8):

$$\overline{e_{ni}^2} = \overline{(e_t + e_n + I_n R_s)^2}$$

$$= \overline{e_t^2} + \overline{e_n^2} + \overline{i_n^2} R_s^2$$

or:

$$e_{Ni}^2 = e_T^2 + e_N^2 + i_N^2 R_s^2 \tag{5.20}$$

or, writing out the source thermal noise e_T in bandwidth Δf specifically:

$$e_{Ni}^2 = 4kTR_s\Delta f + e_N^2 + i_N^2 R_s^2 \tag{5.21}$$

Note that the input noise is dominated by e_N at low values of R_s and by $i_N R_s$ at high values of R_s. In low noise (small e_N, i_N) amplifiers there is a middle range of R_s values where the noise is dominated by source thermal noise. However, if e_N and i_N are high, their ranges of dominance overlap and the source thermal noise is swamped by amplifier noise at all values of the source resistance. This is illustrated in figure 5.6.

5.2.7 Equivalent noise resistance

The amplifier equivalent input noise generators and the total equivalent input noise e_{Ni} may be expressed in terms of their equivalent noise resistances (see section 5.2.5). Thus:

$$R_{nv} = \frac{e_N^2}{4kT\Delta f} \tag{5.22}$$

$$R_{ni} = \frac{4kT\Delta f}{i_N^2} \tag{5.23}$$

$$R_n = \frac{e_{Ni}^2}{4kT\Delta f} \tag{5.24}$$

In general, these resistances are functions of frequency since e_N and i_N are frequency dependent. The broadband equivalent noise resistances are calculated by replacing Δf by the broad bandwidth B. In this case e_N, i_N and e_{Ni} are measured in bandwidth B.

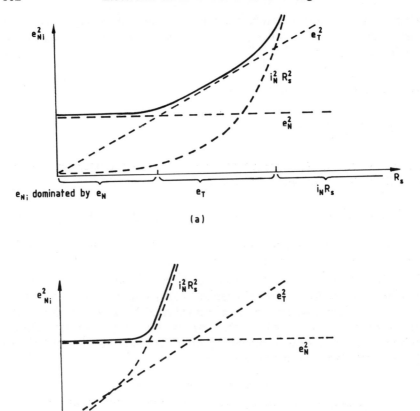

Figure 5.6 Equivalent input mean square noise voltage and noise components versus source resistance and range of component dominance: (a) low noise amplifier; (b) high noise amplifier

5.2.8 Amplifier noise temperature

The noise temperature (T_e) of an amplifier with source resistance R_s is the temperature at which the source resistance has thermal noise equal to the amplifier noise (see section 5.2.3). Thus:

$$4kT_eR_s\Delta f = e_N^2 + i_N^2 R_s^2 \tag{5.25}$$

which gives:

$$T_e = \frac{e_N^2 + i_N^2 R_s^2}{4kR_s\Delta f} \tag{5.26}$$

or in terms of root spectral densities:

$$T_e = \frac{\mathcal{E}_n^2 + \mathcal{I}_n^2 R_s^2}{4kR_s} \tag{5.27}$$

Again, in general, T_e is frequency dependent, and the equations for the broadband T_e are formed by replacing Δf by B.

5.2.9 Noise figure

An ideal amplifier would be noiseless. However, even with such an amplifier the source signal would still be contaminated by the unavoidable thermal noise arising from the source resistance. A useful figure of merit for the amplifier would seem to be one which recognised this and did not indicate merely the closeness to an ideal noiseless amplifier, but rather the degree to which the amplifier added to the noise already present. A commonly used index of noise performance which meets this requirement is the noise factor F, where (Haus *et al.*, 1960a):

$$F = \frac{\text{Total available noise output power}}{\text{Available noise output power arising from thermal noise in source}} \tag{5.28}$$

The source resistance is usually required to be at the standard temperature of 290 K. The terms 'noise factor' and 'noise figure' are often used interchangeably. The convention used here is that F is the noise factor and the noise figure, NF, is F expressed in dB:

$$NF = 10\log_{10}(F) \quad \text{dB} \tag{5.29}$$

The noise figure expresses the fractional increase in noise (from the amplifier) over the unavoidable thermal noise from the source. 'Good' (low noise) amplifiers will have low noise figures. If the amplifier contributes no noise then:

$$NF = 0 \quad \text{dB} \tag{5.30}$$

With the amplifier equivalent noise sources at the input, the above ratio can be calculated at the input since the amplifer gain is the same for both total noise and source thermal noise only. Thus, from equation (5.21):

$$F = e_{Ni}^2/e_T^2 \tag{5.31}$$

$$= 1 + \frac{e_N^2 + i_N^2 R_s^2}{4kTR_s\Delta f} \tag{5.32}$$

Note that F can be expressed in terms of the input and output signal-to-noise ratio (SNR). Thus:

$$F = \frac{e_s^2/e_T^2}{e_s^2 G/(e_{Ni}^2 G)} = \frac{\text{Input SNR}}{\text{Output SNR}} \tag{5.33}$$

where G is the power gain.

In terms of root spectral densities, equation (5.32) can be written:

$$F = 1 + \frac{\mathcal{E}_n^2 + \mathcal{I}_n^2 R_s^2}{4kTR_s} \tag{5.34}$$

The above are spot frequency noise factors. The broadband noise factor is given by:

$$F = 1 + \frac{e_N^2 + i_N^2 R_s^2}{4kTR_s B} \tag{5.35}$$

where e_N and i_N are measured in bandwidth B.

Using equation (5.26), the noise factor can be expressed in terms of the equivalent noise temperature of the amplifier:

$$F = 1 + T_e/T$$

or:

$$F = 1 + T_e/290 \tag{5.36}$$

at the standard temperature of 290 K.

Using equations (5.22) and (5.23), F can also be expressed in terms of the equivalent noise resistances of the input noise generators:

$$F = 1 + R_{nv}/R_s + R_s/R_{ni} \tag{5.37}$$

Note that this is the broadband noise factor if R_{nv} and R_{ni} are broadband measurements or the spot frequency noise factor if they are spot frequency measurements.

From equations (5.32), (5.34), (5.35) or (5.37), it is clear that the noise factor tends to infinity both as R_s tends to zero and as R_s tends to infinity. With a very low source resistance the equivalent input noise voltage represents a large increase in noise over the low thermal noise emf from the source. With a high source resistance, the R_s^2 dependency of the noise power arising from the voltage drop across R_s from the amplifier input noise current, compared with the R_s dependency of the thermal noise, means that the former will dominate.

There will clearly be a value (R_{so}) of R_s for which the noise factor is a minimum. This can be found by differentiating equation (5.32), (5.34), (5.35) or (5.37) with respect to R_s and setting the differential to zero.

In terms of the equivalent noise resistances the optimum value of R_s and corresponding noise factor are:

$$R_{so} = (R_{nv}R_{ni})^{1/2} \tag{5.38}$$

and:

$$F_{min} = 1 + 2(R_{nv}/R_{ni})^{1/2} \tag{5.39}$$

In terms of e_N and i_N or $\mathcal{E}_n$ and $\mathcal{I}_n$ the spot frequency values are:

$$R_{so} = e_N/i_N \tag{5.40}$$

$$F_{min} = 1 + \frac{e_N i_N}{2kT\Delta f} \tag{5.41}$$

or:

$$R_{so} = \mathcal{E}_n/\mathcal{I}_n \tag{5.42}$$

and:

$$F_{min} = 1 + \frac{\mathcal{E}_n \mathcal{I}_n}{2kT} \tag{5.43}$$

and the broadband values are:

$$R_{so} = e_N/i_N \tag{5.44}$$

$$F_{min} = 1 + \frac{e_N i_N}{2kTB} \tag{5.45}$$

The noise factor can be expressed in terms of F_{min} and R_{so} by substituting from any of the above pairs of equations into the corresponding equations for F. Thus:

$$F = 1 + \frac{F_{min} - 1}{2}\left(R_s/R_{so} + R_{so}/R_s\right) \tag{5.46}$$

This is clearly useful for finding F at any source resistance if F_{min} and R_{so} are known.

The variation of noise factor against R_s/R_{so} for various values of R_{nv}/R_{ni} is shown in figure 5.7. As e_N and i_N decrease (R_{nv}/R_{ni} decreases), not only does the noise factor decrease but the minimum becomes broader and a wider range of values of R_s becomes acceptable. Note that it is not always necessary to operate at the minimum noise factor. It is necessary only that the noise is at an acceptable level compared with the signal.

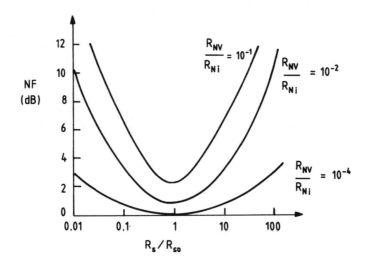

Figure 5.7   Variation of NF with R_s/R_{so}

Note also that R_{so} is not necessarily the value of R_s which gives the maximum power gain (power matching). It is the value of R_s which gives the maximum signal-to-noise ratio, and by analogy with power matching, the use of this value of source resistance is known as noise matching.

In general $\mathcal{E}_n$ and $\mathcal{I}_n$, and therefore the spot frequency values of F, F_{min} and R_{so}, are frequency dependent. The noise factor is therefore a function of both frequency and source resistance. This information may be shown as a variation of NF (or F) against R_s at a number of spot frequencies as shown in figure 5.8, or as contours of the NF versus f and R_s surface, as shown in figure 5.9. In addition, NF_{min} and R_{so} can be shown as functions of frequency, as shown in figure 5.10.

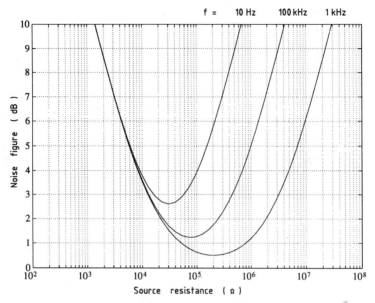

Figure 5.8   Example of the display of noise performance information –
spot frequency noise figure versus R_s

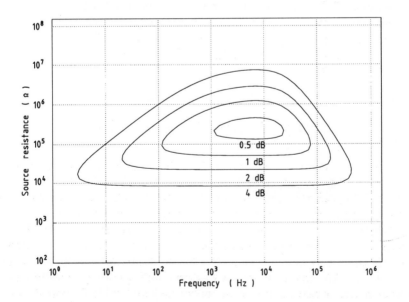

Figure 5.9   Example of the display of noise performance information –
noise figure contours versus source resistance and frequency

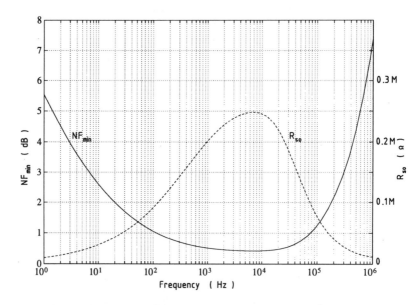

Figure 5.10 Example of the display of noise performance information – variation of R_{so} and NF_{min} with frequency

In the above analyses a purely resistive source has been assumed. If the source has significant series reactive component X_s such that the source impedance is:

$$Z_s = R_s + jX_s$$

as shown in figure 5.11(a), then equation (5.32) becomes:

$$F = 1 + \frac{e_N^2 + i_N^2 |Z_s|^2}{4kTR_s\Delta f}$$

$$= 1 + \frac{e_N^2 + i_N^2 (R_s^2 + X_s^2)}{4kTR_s\Delta f} \tag{5.47}$$

showing that the noise factor is increased by a voltage drop across X_s arising from the amplifier input noise current.

This effect may be removed at one frequency, by connecting a reactance $-X_s$ (a capacitor if X_s is inductive or an inductor if X_s is capacitative) between source and amplifier as shown in figure 5.11(b) (Netzer, 1981). Although the reactances will cancel at only one frequency f_0, the effect of partial cancellation in a frequency band around f_0 may provide a useful reduction in the noise figure over the bandwidth of the signal e_s.

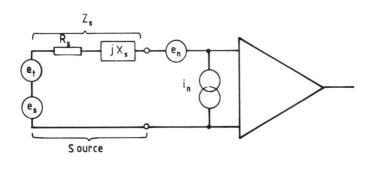

(a)

(b)

Figure 5.11   (a) Source with mixed resistive and reactive source impedance
and (b) cancellation of source reactance

A similar analysis may be carried out by considering the Norton
equivalent circuit of the source (see figure 5.1(b)). In this case, a purely
resistive source may be achieved at one frequency by adding a reactive
component in parallel in order to cancel $B(f)$.

The effect of source reactance when the equivalent input noise generators
are correlated is discussed in section 5.2.14.

5.2.10   Achieving noise matching

Noise matching can be achieved by selecting an amplifier with a value of R_{so}
to suit the source. R_{so} can be changed by altering the operating conditions of
the input semiconductor device (see figure 6.9). Alternatively, the source
resistance can effectively be changed by the use of an input transformer. An
ideal transformer of turns ratio n will transform a source resistance R_s and

its noise voltage $\sqrt{(4kTR_s\Delta f)}$ to a source resistance $n^2 R_s$ and noise voltage $n\sqrt{(4kTR_s\Delta f)}$. Choosing n such that:

$$n^2 R_s = R_{so} \tag{5.48}$$

will achieve noise matching. Since both the source signal and the source thermal noise voltage are transformed by the same amount, there is no change in source SNR by the ideal transformer, but the SNR at the output of the amplifier will be maximised by the noise matching.

In practice, transformers have loss resistances in the windings and cores, and as these are attenuating and thermal noise generating, the SNR and therefore the overall noise factor are degraded. Nevertheless, if the transformer loss resistances referred to the primary can be made small, compared with the source resistance, then noise matching by this method can give a worthwhile improvement in noise figure.

Note that trying to achieve a better SNR by noise matching by adding series or parallel resistors to the source does not work. The added resistors attenuate the source signal and source thermal noise by the same amount, but they attenuate the signal with respect to the amplifier noise and contribute their own thermal noise, so reducing the SNR.

5.2.11 Feedback

The application of feedback to an amplifier cannot change the intrinsic noise-generating processes and the gain change effected by feedback affects signal, source noise and equivalent input amplifier noise alike. The resistive elements of the feedback network contribute their own thermal noise and the input current develops a noise voltage across them, so that the noise figure is degraded. However, if the effect of the feedback components can be made small compared with the source thermal noise, then feedback can be used to alter the input impedance in order to allow power matching, in addition to noise matching or noise matching together with high input impedance to avoid loading the source (Faulkner, 1968; Engberg, 1974).

To illustrate this, we analyse the cases of shunt and series voltage feedback to an amplifier with high gain and input impedance and uncorrelated equivalent input noise generators. In the analysis we assume that low noise resistors are used so that $1/f$ noise is negligible.

An amplifier with shunt feedback is shown in figure 5.12(a). We aim to find the noise sources in the equivalent circuit in figure 5.12(b). Firstly, we find the output resulting from each source acting alone. As in the small signal analysis of feedback applied to high gain amplifiers, we calculate the output voltage that will give a feedback current i, which, added to the source current at the input, will give a zero (in practice very small) amplifier input voltage.

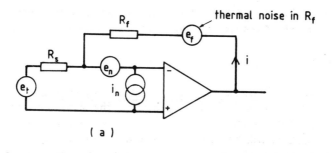

(a)

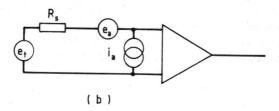

(b)

Figure 5.12 (a) Amplifier with series current feedback and (b) equivalent circuit

As we are interested only in the noise in this analysis, the signal source has been omitted. The voltage gain $K = -R_f/R_s$.

The following calculations are of the output voltage resulting from each noise source acting alone.

Source:

e_n

$$iR_s = -e_n$$

$$\text{Output } |e_{no}| = |iR_sK| = e_nR_f/R_s$$

e_f

$$\text{Output } e_{fo} = -e_f$$

i_n

$$i = -i_n$$

$$\text{Output } |e_{io}| = i_nR_f$$

e_t

$$\text{Output } |e_{to}| = Ke_t$$

Adding the mean squares of these outputs (independent sources) and dividing by K^2 gives the total equivalent input measured noise voltage:

$$e_{Ni}^2 = e_N^2 + 4kTBR_s^2/R_f + i_N^2 R_s^2 + 4kTR_s B$$

We find e_A by setting $R_s = 0$ (see equation (5.21)); then:

$$e_A = e_N \tag{5.49}$$

We find i_A^2 by making R_s very large, dropping small terms and dividing by R_s^2 (see equation (5.21)); then:

$$i_A^2 = i_N^2 + 4kTB/R_f \tag{5.50}$$

Thus the only extra term is a contribution to the equivalent input noise current from thermal noise in the feedback resistor.

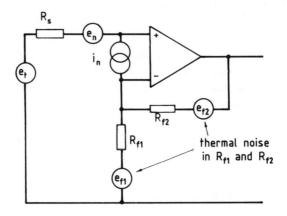

Figure 5.13   Amplifier with series voltage feedback

A similar analysis of the series voltage feedback case shown in figure 5.13 leads to:

$$i_A = i_N \tag{5.51}$$

and:

$$e_A^2 = e_N^2 + 4kTR_p B + (i_N R_p)^2 \tag{5.52}$$

where:

$$R_p = \frac{R_{f_1} R_{f_2}}{R_{f_1} + R_{f_2}} \qquad (5.53)$$

Thus the feedback contributes terms in e_A^2 resulting from thermal noise from, and i_N flowing in, the feedback resistors. So far as the noise analysis is concerned, the input circuit behaves as if R_s has been increased by R_p.

5.2.12 Cascaded, matched networks

Consider an amplifier which is power-matched to the source and load as shown in figure 5.14.

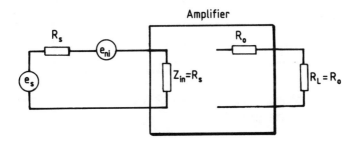

Figure 5.14 Matched amplifier

From equation (4.11), the available thermal noise power of the source in frquency band Δf is:

$$P_t = kT\Delta f \qquad (5.54)$$

The available output power from this source is :

$$P_{ot} = G_a P_t \qquad (5.55)$$

where G_a is the available power gain of the amplifier.

The actual available output noise power (from the definition of F – equation (5.28)) is:

$$P_{on} = FG_a P_t \qquad (5.56)$$

Therefore the available output noise power added by the amplifier is:

$$P_{oa} = (F - 1)G_a P_t \qquad (5.57)$$

or, referred to the input:

$$P_{na} = (F - 1)P_t$$

$$= (F - 1)kT\Delta f \tag{5.58}$$

Therefore, we can show the amplifer as an ideal amplifier with inputs of available signal power (P_s), available source noise power ($kT\Delta f$) and added amplifier noise power (($(F - 1)kT\Delta f$)), as shown in figure 5.15.

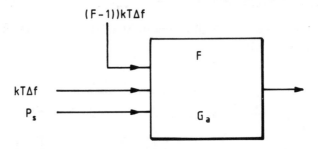

Figure 5.15 Matched amplifier model with noise factor F and available power gain G_a

Consider two amplifiers in cascade with the input of the second, power-matched to the first (figure 5.16). The noise factor for the pair is (adding all the contributions at the output and dividing by the overall gain $G_{a_1} G_{a_2}$):

$$F = \frac{\text{Total output noise power}}{\text{Noise output power from source}}$$

$$= F_1 + \frac{(F_2 - 1)}{G_{a_1}} \tag{5.59}$$

This analysis may be extended to more amplifiers leading to:

$$F = F_1 + \frac{(F_2 - 1)}{G_{a_1}} + \frac{(F_3 - 1)}{G_{a_1} G_{a_2}} \cdots \tag{5.60}$$

This is known as Friis's formula (Friis, 1944).

Provided that the gain of the first amplifier is reasonably high, and the noise figures of subsequent amplifiers are not too high, the overall noise figure is dominated by that of the first amplifier.

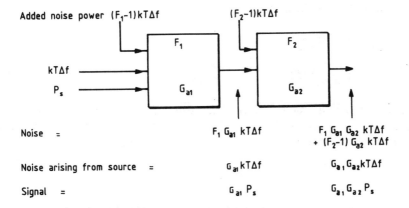

Figure 5.16 Amplifiers in cascade

Since F may be expressed in terms of noise temperature, we have (from (5.36) and (5.60)):

$$T_e = T_{e_1} + \frac{T_{e_2}}{G_{a_1}} + \frac{T_{e_3}}{G_{a_1} G_{a_2}} \cdots \tag{5.61}$$

5.2.13 Noise figure of matched attenuator

If we connect two resistors in series or parallel, then the thermal noise generated is that of a single resistance having the value appropriate to the connection (resistances or conductances added for a series or parallel connection respectively). This can be extended to further resistances and leads to the result that the thermal noise from the port of a passive network is that from a resistance having the same value as the output resistance of the port.

If we consider a source resistance connected to a matched attenuator, the output port of the attenuator (the attenuator plus source being a passive network) has the same resistance and therefore the same available noise power as the source. That is:

$$P_{on} = P_t$$

and from equation (5.56):

$$F = 1/G_a$$

or:

$$NF = 10\log(1/G_a)$$

where G_a is the 'gain' ($\leqslant 1$) of the attenuator.

But the loss of an attenuator is defined as:

$$L = -10\log(G_a) = 10\log(1/G_a) \quad \text{dB}$$

therefore:

$$NF = L \quad \text{dB} \tag{5.62}$$

5.2.14 Two-port network with correlated equivalent noise generators

In general, the two equivalent input noise generators of a noisy two-port network are correlated. The degree of correlation is often significant in high frequency amplifiers as a result of capacitative coupling – for example, via the base/collector capacitance in a bipolar transistor.

In order to provide easy comparison with other published work, a somewhat different approach will be adopted in this analysis (following Rothe and Dahlke, 1956; Haus *et al.*, 1960b), in that it will be in terms of noise current generators and admittances rather than voltage generators and impedances. In addition, as will be seen, there is an implicit requirement for the phase relationship between the frequency components of the correlated noise signals to be maintained in the analysis. For that reason, the analysis is carried out using the Fourier transforms of noise signals of duration τ ($\tau \to \infty$). It follows from equation (2.74) that the relationships between a time signal ($i(t)$), its Fourier transform $I(f)$ and its normalised power spectrum $S(f)$ are:

$$I(f) = \mathcal{F}[i(t)]$$

$$S(f) = \lim_{\tau \to \infty} \left[\frac{2\overline{|I(f)|^2}}{\tau} \right] \tag{5.63}$$

A source with admittance:

$$Y_s = G_s + jB_s \tag{5.64}$$

where G_s is the source conductance, B_s the source susceptance and thermal noise i_t connected to a noisy amplifier, is shown in figure 5.17(a). For convenience the generators have been labelled by the Fourier transform of their time functions.

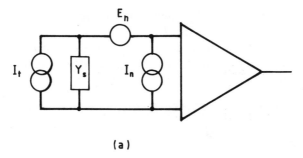

(a)

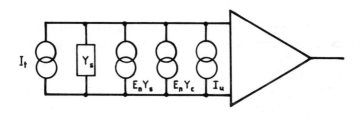

(b)

Figure 5.17 (a) Amplifier driven by source with admittance Y_s and
(b) all-current-generator equivalent circuit

Firstly, the voltage generator is converted to its equivalent current generator $Y_s E_n$. Secondly, the current i_n which is partially correlated with e_t is split into an uncorrelated current i_u and a fully correlated current i_c. Then:

$$i_n = i_c + i_u$$

and:

$$I_n = I_c + I_u \tag{5.65}$$

Since I_c is fully correlated with E_c, they are proportional. The constant of proportionality has the form of an admittance and is called the correlation admittance $Y_c = G_c + jB_c$. Thus:

$$I_c = E_n Y_c \tag{5.66}$$

The circuit now has the form shown in figure 5.17(b).

Combining generators $E_n Y_s$ and $E_n Y_c$, we are left with three uncorrelated generators I_t, $(Y_c + Y_s)E_n$ and I_u.

The noise factor is the ratio of the normalised power spectrum of the current in the input circuit to that resulting from I_t only. Thus, noting the result in equation (5.63):

$$F = \frac{\overline{|I_t|}^2 + |Y_s + Y_c|^2 \overline{|E_n|}^2 + \overline{|I_u|}^2}{\overline{|I_t|}^2}$$

$$= \frac{S_{ti}(f) + ((G_s + G_c)^2 + (B_s + B_c)^2)S_{nv}(f) + S_u(f)}{S_{ti}(f)} \tag{5.67}$$

where $S_{ti}(f)$, $S_{nv}(f)$ and $S_u(f)$ are the normalised spectral densities of i_t, e_n and i_u respectively.

Replacing the spectral densities by their equivalent noise conductances (reciprocal of their equivalent noise resistances – see section 5.2.5), that is:

$$S_{ti}(f) = 4kTG_s$$

$$S_{nv}(f) = 4kT/G_{nv}$$

$$S_u(f) = 4kTG_u$$

then:

$$F = 1 + \frac{G_u + ((G_s + G_c)^2 + (B_s + B_c)^2)/G_{nv}}{G_s} \tag{5.68}$$

Note that this equation involves the equivalent noise conductance of the uncorrelated part of the equivalent input noise current. It can be expressed in terms of the equivalent noise conductance (G_{ni}) of the total input noise current by noting that, from equations (5.65) and (5.66), and bearing in mind that I_c and I_u are by definition uncorrelated:

$$\overline{|I_n|}^2 = |Y_c|^2 \overline{|E_n|}^2 + \overline{|I_u|}^2 \tag{5.69}$$

Using this to substitute for $\overline{|I_u|}^2$ in equation (5.67), leads to:

$$F = 1 + \frac{G_{ni} + ((G_s + G_c)^2 + (B_s + B_c)^2 - G_c^2 - B_c^2)/G_{nv}}{G_s} \tag{5.70}$$

As expected, if B_s, G_c and B_c are put equal to zero, the above reduces to equation (5.37).

From equations (5.68) or (5.70) it can be seen that the noise factor is now a function of source admittance – that is, both source conductance and susceptance. The minimum value of F and the source admittance at which it occurs may be found by setting to zero the partial differentials of F with respect to G_s and B_s. The optimum values of the source conductance (G_{so}) and susceptance (B_{so}) at which F is a minimum (F_{min}) are:

$$G_{so} = (G_u G_{nv} + G_c^2)^{1/2} \qquad (5.71)$$

or:

$$G_{so} = (G_{ni} G_{nv} - B_c^2)^{1/2} \qquad (5.72)$$

$$B_{so} = -B_c \qquad (5.73)$$

and:

$$F_{min} = 1 + 2(G_{so} + G_c)/G_{nv} \qquad (5.74)$$

The noise factor may now be written:

$$F = F_{min} + \frac{(G_s - G_{so})^2 + (B_s - B_{so})^2}{G_{nv} G_s} \qquad (5.75)$$

or:

$$F = F_{min} + \frac{|Y_s - Y_{so}|^2}{G_{nv} G_s} \qquad (5.76)$$

To obtain noise matching (F_{min}), in this case, both source conductance and susceptance must be matched. It follows from equation (5.73) that the source susceptance should be the same magnitude as the correlation susceptance and inductive if B_c is capacitative, and capacitative if B_c is inductive. The optimum source conductance (equation (5.71)) is governed not only by e_n and i_n but also by their degree of correlation. Since, in general, Y_c, G_u and G_{nv} are frequency dependent, a match may only be possible at one frequency, although the lower the noise level of the amplifier, the broader the minimum and the wider the frequency band over which the noise factor may be acceptable.

The number of variables involved leads to a difficulty in presenting the variation of noise factor and matching conditions. For example, a graph of F against G_s and B_s is a surface which may be represented by contour lines, as shown in the example in figure 5.18, but we need one plot like this for each of the spot frequencies of interest.

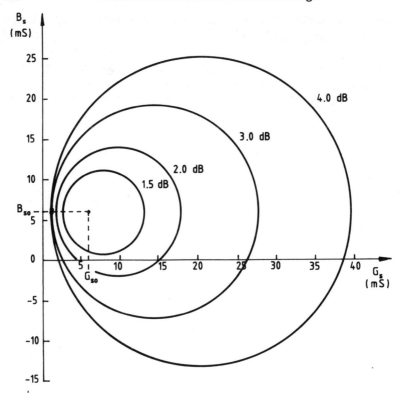

Figure 5.18   Noise figure contours versus source conductance (G_s)
and susceptance (B_s) for a device with $NF_{min} = 0.95$ dB,
$G_{nv} = 22.5$ mS, $G_{so} = 6.0$ mS and $B_{so} = 6.0$ mS

However, from equation (5.75) it can be seen that if F_{min}, G_{so}, B_{so} and G_{nv} are shown as functions of frequency, then F may be calculated for any Y_s at any frequency.

Noise matching at a particular frequency does not guarantee resonance at that frequency. To achieve noise matching, the susceptance of the source must be equal and opposite to the correlation susceptance. For resonance, it should be equal and opposite to the amplifier input susceptance at the resonant frequency.

References

Becking, A. G., Groendijk, H. and Knol, K. S. (1955). 'The noise factor of four-terminal networks', *Philips Research Reports*, **10**, 349–357.

Engberg, J. (1974). 'Simultaneous input power match and noise optimization using feedback', in *Proceedings of the 4th European Microwave Conference, Montreux, Switzerland*, pp. 385–389.

Faulkner, E. A. (1968). 'The design of low-noise audio-frequency amplifiers', *The Radio and Electronic Engineer*, **36(1)**, 17–30.

Friis, H. T. (1944). 'Noise figure of radio receivers', *Proceedings of the IRE*, **32**, 419–422.

Haus, H. A. *et al.* (1960a). 'IRE standards on methods of measuring noise in linear twoports, 1959', *Proceedings of the IRE*, **48**, 60–68.

Haus, H. A. *et al.* (1960b). 'Representation of noise in linear twoports', *Proceedings of the IRE*, **48**, 69–74.

Letzter, S. and Webster, N. (1970). 'Noise in amplifiers', *IEEE Spectrum*, **7(8)**, 67–75.

Montgomery, H. C. (1952). 'Transistor noise in circuit applications', *Proceedings of the IRE*, **40**, 1461–1471.

Netzer, Y. (1981). 'The design of low-noise amplifiers', *Proceedings of the IEEE*, **69(6)**, 728–742.

Nyquist, H. (1928). 'Thermal agitation of electric charge in conductors', *Physics Review*, **32**, 110–113.

Rothe, H. and Dahlke, W. (1956). 'Theory of noisy fourpoles', *Proceedings of the IRE*, **44**, 811–818.

6   Noise Models

6.1 Introduction

In order to analyse the noise performance of circuits it is useful to represent real devices by 'ideal' noiseless circuits, plus a minimum number of noise generators. In the following sections we develop the noise models of some common circuit elements.

6.2 Resistor

The resistor generates thermal noise and $1/f$ noise. The voltage and current generator based models are shown in figure 6.1. The shunt capacitance and series inductance – important at high frequencies – are not shown. These stray reactive components will alter the noise spectral density significantly only at frequencies sufficiently high that their reactances are comparable with R.

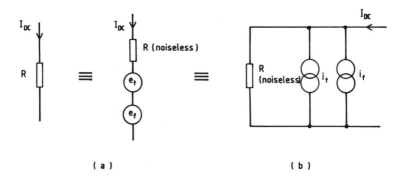

Figure 6.1   Noise models of resistor (a) series voltage generators;
(b) parallel current generators

The spectral densities of e_t and i_t (thermal noise) are given by equations (4.1) and (4.4) respectively. The spectral densities of e_f and i_f ($1/f$ noise) are (from equations (4.27), (4.30) and (4.34)):

$$S_f(f) = 0.434 C_F^2 (I_{DC}R)^2 f^{-1} \tag{6.1}$$

$$S_{fi}(f) = 0.434 C_F^2 I_{DC}^2 f^{-1} \tag{6.2}$$

when α is approximately equal to 1.

The constant C_F depends on resistor type, being high in carbon composition and low in wire-wound and metal-film resistors. In low noise applications, resistors with low excess noise are obviously the preferred choice, together with low direct current.

6.3 Capacitor

The ideal capacitor is noiseless. The real capacitor has resistive components associated with it, as shown in figure 6.2. R_p is the leakage resistance and R_s is the sum of the lead resistance and the resistance representing dielectric loss. This latter component increases with frequency. There is thermal (e_{ts}) and flicker (e_{fs}) noise associated with R_s and thermal noise (e_{tp}) with R_p. In practice, their noise is normally quite negligible compared with that for other circuit components.

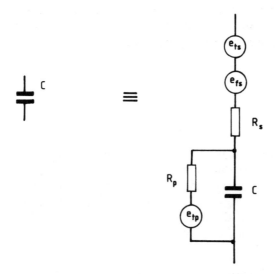

Figure 6.2   Capacitor noise model

Note that the series lead inductance has not been included. When used within a frequency range appropriate for each type of capacitor, this stray reactance is usually negligible.

6.4 Inductor

Again, the ideal inductor is noiseless. The real inductor, shown in figure 6.3, has a series resistance which is the sum of the wire resistance and a resistance representing the effect of hysteresis and eddy current loss in the core. The loss resistance increases with frequency. Note that the core losses may also be represented by a resistor in parallel with L. There is thermal noise associated with R_s and flicker noise (normally negligibly small) associated with the wire resistance component if direct current is flowing through the inductor.

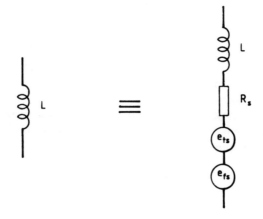

Figure 6.3   Inductor noise model

6.5 Transformer

The transformer has the same parasitic resistances as the inductor. The equivalent circuit is shown in figure 6.4. The resistances R_{ps} and R_{ss} are respectively the primary and secondary winding resistances, R_{cl} represents the core losses and L_p is the primary inductance. The resistive components all have associated thermal noise. If direct current flows through the windings, R_{ps} and R_{ss} will have associated flicker noise (normally negligibly small).

6.6 Antenna

Treated as a one-port network, the antenna has the equivalent circuit as shown in figure 6.5.

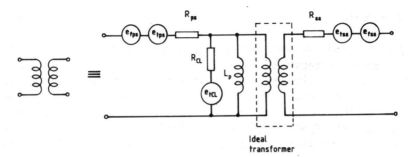

Figure 6.4 Transformer noise model

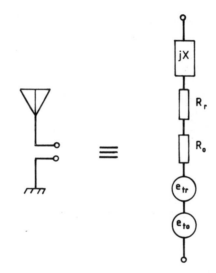

Figure 6.5 Antenna noise model

The resistive part of the series impedance has two components. R_o represents the dissipative resistance of the antenna and R_r its radiation resistance. Both are frequency dependent. R_o is usually very small compared with R_r and is significant only in antennas whose dimensions are very small compared with a wavelength (Collin and Zucker, 1969). The thermal noise associated with R_o is e_{to}. The noise e_{tr} associated with the radiation resistance R_r is not thermal noise determined by the physical temperature of the antenna since it is not in thermal equilibrium with its immediate surroundings. The noise is induced by electromagnetic radiation from a number of sources, including electrical machinery, atmospheric electrical activity and cosmic radiation. This noise depends not only on frequency but also on the orientation of the antenna, its directivity pattern, and time

(Rotkiewicz, 1982). The equivalent temperature is often used in connection with antennas, and at frequencies above approximately 100 MHz this temperature may be significantly below the ambient temperature. It follows from chapter 5, section 5.2.3 that, if $R_o \ll R_r$:

$$T_e = S_{tr}(f)/(4kR_r(f)) \tag{6.3}$$

where $S_{tr}(f)$ is the normalised power spectrum of e_{tr}.

6.7 Semiconductor diode

The current flowing through a semiconductor diode is given by:

$$I = I_0 \left(\exp\left[\frac{eV}{kT}\right] - 1 \right) \tag{6.4}$$

where V is the voltage drop across the diode and I_0 is the saturation reverse bias leakage current.

The noise behaviour can be modelled as if this current can be split into two currents:

$$I_1 = -I_0$$

and:

$$I_2 = I_0 \exp\left[\frac{eV}{kT}\right]$$

each giving rise to independent shot noise currents (van der Ziel and Chenette, 1978; Buckingham, 1983).

Thus, when I_1 and I_2 are comparable, the total mean square noise current in frequency band Δf is:

$$i_N^2 = i_{N_1}^2 + i_{N_2}^2$$

where:

$$i_{N_1}^2 = 2e|I_1|\Delta f$$

and:

$$i_{N_2}^2 = 2e|I_2|\Delta f$$

At $V=0$, $I_1=-I_2$ and:

$$i_N^2 = 4eI_0\Delta f \tag{6.5}$$

For a strongly forward biased diode:

$$I_2 \gg I_1$$

and:

$$i_N^2 = 2e\Delta f I_0 \exp\left[\frac{eV}{kT}\right]$$

In addition, there is flicker noise associated with I, with a mean square value of:

$$i_F^2 = \frac{K_F I'\Delta f}{f^\alpha} \tag{6.6}$$

The small signal AC resistance (r_e) of a forward biased diode is obtained by differentiating equation (6.4) giving:

$$r_e = \frac{kT}{eI} \tag{6.7}$$

$$= 0.026I^{-1} \quad \Omega \text{ at } 25°C$$

The noise model of a strongly forward biased diode is shown in figure 6.6. The parasitic series resistance R_s and associated thermal noise generator are shown. Both are usually small.

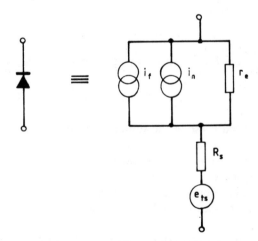

Figure 6.6 Noise model of forward biased diode

6.8 Bipolar transistor

The noise performance of the bipolar transistor may be analysed using either the T (Nielsen, 1957; van der Ziel and Chenette, 1978) or hybrid-π (Baxandall, 1968) equivalent circuits. We arbitrarily choose the latter for the analysis presented here.

The conventional hybrid-π small signal model of the bipolar transistor together with the dominant noise sources is shown in figure 6.7.

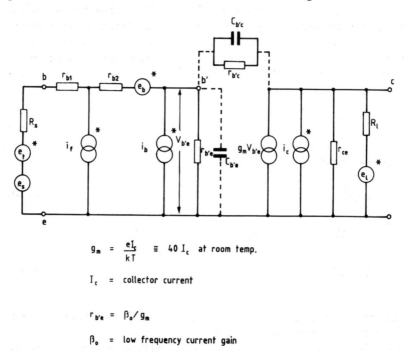

$$g_m = \frac{e I_c}{kT} \cong 40\, I_c \text{ at room temp.}$$

I_c = collector current

$r_{b'e} = \beta_o / g_m$

β_o = low frequency current gain

Figure 6.7   Hybrid-π transistor circuit with noise sources

The noise generators are indicated by an asterisk. They are:

e_t – thermal noise in the source resistance R_s
e_b – thermal noise in the base resistance $r_{bb'}$
i_b – shot noise in the base current I_B
i_c – shot noise in the collector current I_C
e_l – thermal noise in the load resistor R_l
i_f – flicker noise in the base current.

Note that the base resistance $r_{bb'}$ has been split into two resistors r_{b_1} and r_{b_2}. It has been found that the mechanisms causing flicker noise in the base current do not involve the whole base region and connection to the same extent (Jaeger and Brodersen, 1970). The effect may be modelled as shown. The two resistances (r_{b_1} and r_{b_2}) are roughly equal for planar transistors (Motchenbacher and Fitchen, 1973).

There is no thermal noise associated with $r_{b'e}$ and r_{ce} as these are not real, dissipative resistors, but the slopes of the $V - I$ curves at those points in the circuits.

We shall, at first, restrict our analysis to frequencies below those at which the capacitors $C_{b'e}$ and $C_{b'c}$ have a significant effect, where the gain of the transistor amplifier is sufficiently high that thermal noise in the load can be neglected, compared with the amplified equivalent input noise, and to transistors and operating conditions where the feedback resistance $r_{b'c}$ is sufficiently high for its effect to be negligible. The mean square noise generator outputs are:

$$e_T^2 = 4kTR_s\Delta f \tag{6.8}$$

$$e_B^2 = 4kTr_{bb'}\Delta f \tag{6.9}$$

$$i_B^2 = 2eI_B\Delta f \tag{6.10}$$

$$i_C^2 = 2eI_C\Delta f \tag{6.11}$$

$$i_F^2 = \frac{K_F I_B^\gamma \Delta f}{f^\alpha} \tag{6.12}$$

where α and γ are usually between 1 and 2, and K_F (the flicker coefficient) varies widely between types and samples.

With reference to figure 6.7, we now form Thévenin equivalents to the Norton current generators formed by i_f, i_b and the parallel resistances to their left. These voltage generators are:

$$e_{bf} = i_f(R_s + r_{b1})$$

and:

$$e_{ib} = i_b(R_s + r_{bb'}) \tag{6.13}$$

respectively.

We now insert a voltage generator (e_{ic}) in the base circuit which gives the same output current as i_c. This requires a generator which, acting alone, develops a voltage $v_{b'e}$ across $r_{b'e}$ such that $g_m v_{b'e} = i_c$. Thus:

$$e_{ic}r_{b'e}(R_s + r_{bb'} + r_{b'e})^{-1} = i_c/g_m$$

or:

$$e_{ic} = i_c(R_s + r_{bb'} + r_{b'e})(g_m r_{b'e})^{-1} \tag{6.14}$$

We now have the circuit shown figure 6.8(a).

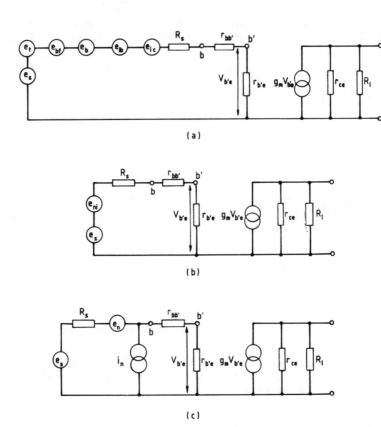

Figure 6.8   Equivalents to noisy hybrid-π circuit

Combining the (independent) noise generators we have the circuit in figure 6.8(b) where:

$$e_{Ni}^2 = e_T^2 + e_B^2 + e_{Ib}^2 + e_{Ic}^2 + e_{Bf}^2 \tag{6.15}$$

and substituting from equations (6.8)–(6.14):

$$e_{Ni}^2 = 4kT\Delta f(R_s + r_{bb'}) + 2eI_B\Delta f(R_s + r_{bb'})^2$$

$$+ 2eI_C\Delta f(R_s + r_{bb'} + r_{b'e})^2 g_m^{-2} r_{b'e}^{-2} + K_F I_B^\nu f^{-\alpha}(R_s + r_{b_1})^2 \Delta f \quad (6.16)$$

We obtain the equivalent series noise generator (e_n) (figure 6.8(c)) mean square voltage, as in the case of the general amplifier, by putting R_s equal to zero (see equation (5.21)). Then:

$$e_N^2 = 4kT\Delta f r_{bb'} + 2eI_B\Delta f r_{bb'}^2 + 2eI_C\Delta f(r_{bb'} + r_{b'e})^2 g_m^{-2} r_{b'e}^{-2}$$

$$+ K_F I_B^\nu r_{b_1}^2 \Delta f f^{-\alpha} \quad (6.17)$$

and the root spectral density is:

$$\mathcal{E}_n = \left(4kT r_{bb'} + 2eI_B r_{bb'}^2 + 2eI_C(r_{bb'} + r_{b'e})^2 g_m^{-2} r_{b'e}^{-2} + K_F I_B^\nu r_{b_1}^2 f^{-\alpha}\right)^{1/2}$$

$$(6.18)$$

We obtain i_N^2(from (6.16)) by letting R_s increase, dropping small terms (keeping only terms involving R_s^2) and dividing by R_s^2 (see equation (5.21)). Then:

$$i_N^2 = 2eI_B\Delta f + 2eI_C\Delta f g_m^{-2} r_{b'e}^{-2} + K_F I_B^\nu \Delta f f^{-\alpha} \quad (6.19)$$

and:

$$\mathcal{I}_n = \left(2eI_B + 2eI_C g_m^{-2} r_{b'e}^{-2} + K_F I_B^\nu f^{-\alpha}\right)^{1/2} \quad (6.20)$$

We now simplify these equations by restricting the consideration to low noise transistors with $I_c \leqslant 1$ mA and $\beta_0 \geqslant 100$. In this case since:

$$r_{b'e} = \beta_0/g_m$$

$$\simeq \beta_0/(40 I_C)$$

$$> 2.5 \times 10^3 \ \Omega \quad (6.21)$$

and $r_{bb'} \sim 200\,\Omega$ or less for low noise transistors we can neglect $r_{bb'}$ compared with $r_{b'e}$ and (6.17) and (6.18) become:

$$e_N^2 = 4kT\Delta f r_{bb'} + 2e\Delta f(I_B r_{bb'}^2 + I_C g_m^{-2}) + K_F I_B^\nu r_{b_1}^2 \Delta f f^{-\alpha} \quad (6.22)$$

and:

$$\mathcal{E}_n = \left(4kTr_{bb'} + 2e(I_B r_{bb'}^2 + I_C g_m^{-2}) + K_F I_B'' r_{b_1}^2 f^{-\alpha}\right)^{1/2} \qquad (6.23)$$

If we restrict I_C to less than $0.3\,\text{mA}$ we can see, by putting $g_m \sim 40 I_C$, $r_{bb'} \sim 200\,\Omega$ and $I_C/I_B > 50$ that:

$$I_B r_{bb'}^2 \ll I_C/g_m^2 \qquad (6.24)$$

and we can write:

$$e_N^2 \simeq 4kT\Delta f r_{bb'} + 2e\Delta f I_C g_m^{-2} + K_F I_B'' r_{b_1}^2 \Delta f f^{-\alpha} \qquad (6.25)$$

and:

$$\mathcal{E}_n \simeq \left(4kTr_{bb'} + 2eI_C g_m^{-2} + K_F I_B'' r_{b_1}^2 f^{-\alpha}\right)^{1/2} \qquad (6.26)$$

Since $g_m r_{b'e} = \beta_0$ we can see that the second term on the right of equation (6.19) can be neglected compared with the first and:

$$i_N^2 \simeq 2eI_B\Delta f + K_F I'' \Delta f f^{-\alpha} \qquad (6.27)$$

and:

$$\mathcal{I}_n = \left(2eI_B + K_F I_B'' f^{-\alpha}\right)^{1/2} \qquad (6.28)$$

At frequencies sufficiently high for the flicker noise terms to be negligible, (6.25)–(6.28) become:

$$e_N^2 \simeq 4kT\Delta f r_{bb'} + 2e\Delta f I_C g_m^{-2} \qquad (6.29)$$

$$\mathcal{E}_n \simeq \left(4kTr_{bb'} + 2eI_C g_m^{-2}\right)^{1/2} \qquad (6.30)$$

$$i_N^2 \simeq 2eI_B\Delta f \qquad (6.31)$$

and:

$$\mathcal{I}_n = \left(2eI_B\right)^{1/2} \qquad (6.32)$$

Note that the sources of the noise in i_n and e_n are different, and therefore i_n and e_n are uncorrelated under these conditions.

The equivalent noise resistances of these generators are (from (5.22) and (5.23)):

$$R_{nv} = r_{bb'} + (2g_m)^{-1} \tag{6.33}$$

$$R_{ni} = 2\beta_{DC}g_m^{-1} \tag{6.34}$$

where we have used:

$$g_m = eI_C(kT)^{-1} \tag{6.35}$$

and:

$$\beta_{DC} = I_C/I_B \tag{6.36}$$

We now use these quantities to calculate the optimum source resistance R_{so} and the minimum noise figure F_{min} (from (5.38) and (5.39)). Thus:

$$R_{so} = (2\beta_{DC}r_{bb'}g_m^{-1} + \beta_{DC}g_m^{-2})^{1/2} \tag{6.37}$$

and:

$$F_{min} = 1 + (2r_{bb'}g_m\beta_{DC}^{-1} + \beta_{DC}^{-1})^{1/2} \tag{6.38}$$

As I_C decreases, g_m ($\simeq 40I_C$) decreases and the limit of F_{min} is:

$$F_{min} = 1 + \beta_{DC}^{-1/2} \tag{6.39}$$

at:

$$R_{so} = \beta_{DC}^{1/2}g_m^{-1}$$

$$= 0.025\beta_{DC}^{1/2} I_C^{-1} \tag{6.40}$$

The variation of $\mathcal{E}_n, \mathcal{I}_n, F_{min}$ and R_{so} with collector current in the mid-frequency region is shown in figure 6.9. Note that these curves should be taken as a rough guide only, since not only does their validity depend on the assumptions specifically made above, but also that β_{DC} and $r_{bb'}$ are independent of collector current.

From these equations and graphs it is clear that for low noise operation in the mid-frequency region we need a low $r_{bb'}$, high β_{DC} and operation at low I_C. It follows from (6.38) and (6.39) that, if these conditions are met then F_{min} is close to unity and varies slowly with I_C. On the other hand, since g_m is proportional to I_C, it follows that R_{so} may be varied by altering I_C.

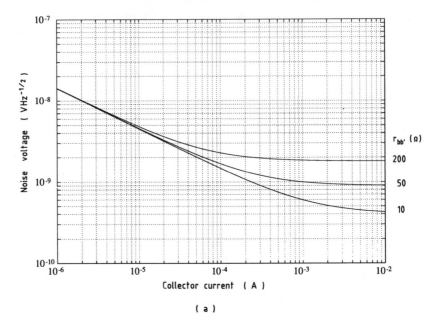

(a)

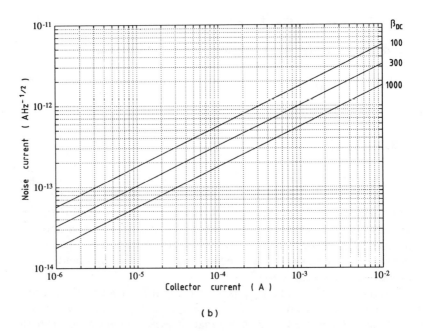

(b)

Figure 6.9 Variation with collector current of: (a) $\mathcal{E}_n$, (b) $\mathcal{I}_n$, (c) F_{min} and (d) R_{so}, in mid-frequency region

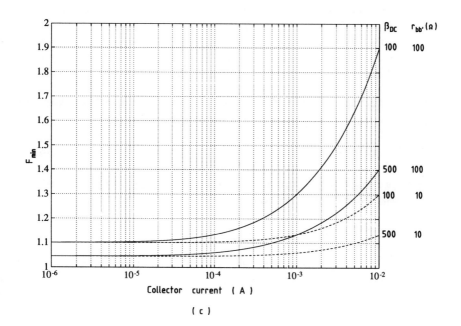

(c)

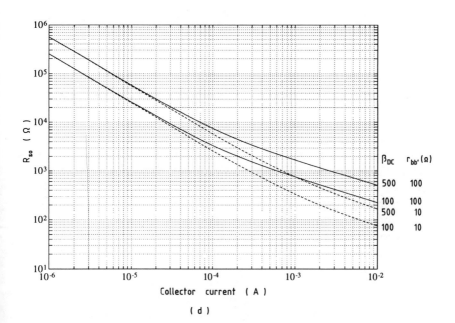

(d)

If we go back and examine the flicker noise terms it is clear that again for low noise operation we need low I_B (that is, low I_C and high β_{DC}). It should be noted that the flicker noise terms in $\mathcal{E}_n$ and $\mathcal{I}_n$ are from the same source and we would expect e_n and i_n to be partially correlated at low frequencies.

The flicker noise corner frequencies – the frequencies at which the contributions of flicker and constant spectral density noise are equal – are different for the two equivalent noise generators. From (6.28), taking the case of α equal to unity, we have for the $\mathcal{I}_n$ corner frequency:

$$f_{Li} = K_F I_B^{n-1} (2e)^{-1} \tag{6.41}$$

Similarly, from (6.26) (taking $r_{b_1} = r_{bb'}$), the corner frequency for the equivalent input voltage generator is:

$$f_{Lv} = K_F I_B^n r_{bb'}^2 (4kT r_{bb'} + 2e I_C g_m^{-2})^{-1} \tag{6.42}$$

If $r_{bb'}$ is sufficiently small for the thermal noise term to be neglected compared with the shot noise term then:

$$f_{Lv} = K_F I_B^n r_{bb'}^2 g_m^2 (2e I_C)^{-1}$$

$$= K_F I_B^{n-1} (2e)^{-1} (1600 \, r_{bb'}^2 \, I_C^2 \, \beta_{DC}^{-1})$$

$$= f_{Li} (1600 \, r_{bb'}^2 \, I_C^2 \, \beta_{DC}^{-1}) \tag{6.43}$$

For low I_C, the term in brackets is usually small, leading to a small f_{Lv} compared with f_{Li}. For example, if I_C is 100 μA, $r_{bb'}$ is 200 Ω and β_{DC} is 100, then f_{Lv}/f_{Li} is 6.4×10^{-3}. Note that the terms we have neglected would make f_{Lv}/f_{Li} even smaller.

At high frequencies, when the gain of the transistor is reduced by the shunting of $r_{b'e}$ by the reactance of $C_{b'e}$ and feedback through $C_{b'c}$, terms that we have dropped will become significant. The collector current shot noise will have an increasing contribution to $\mathcal{I}_n$, as will shot noise currents flowing through the base resistance contribute to $\mathcal{E}_n$. Significant correlation of e_n and i_n at higher frequencies requires account to be taken of the correlation admittance when calculating the noise factor (see chapter 5, section 5.2.14).

A crude approximation, which should not be used to compute noise levels but which nevertheless serves to illustrate the difference in high frequency behaviour of $\mathcal{E}_n$ and $\mathcal{I}_n$, is to ignore the feedback capacitor $C_{b'c}$ but to include the larger capacitance $C_{b'e}$. In this case, $r_{b'e}$ may be replaced in the equations by $Z_{b'e}(Z_{b'e} = r_{b'e}//X_{cb'e})$ and $r_{b'e}^2$ by $|Z_{b'e}|^2$. Then (6.19) and (6.20), ignoring flicker noise, become:

$$i_N^2 = 2eI_B\Delta f + 2eI_C\Delta f(g_m|Z_{b'e}|)^{-2} \tag{6.44}$$

and:

$$\mathcal{I}_n = \left(2eI_B + 2eI_C(g_m|Z_{b'e}|)^{-2}\right)^{1/2} \tag{6.45}$$

At low frequencies, the second term is negligible for the reasons used in the derivation of (6.27). The term becomes significant only at low values of $X_{cb'e}$. Then $Z_{b'e}$ may be replaced by $X_{cb'e}$ and:

$$i_N^2 = 2eI_B\Delta f\left((1 + \beta_{DC}(f/f_T)^2\right) \tag{6.46}$$

or:

$$\mathcal{I}_n = \left(2eI_B\left(1 + \beta_{DC}(f/f_T)^2\right)\right)^{1/2} \tag{6.47}$$

where:

$$f_T = g_m(2\pi C_{b'e})^{-1} \tag{6.48}$$

The resistance $r_{b'e}$ (to be replaced by $Z_{b'e}$) does not appear in the approximation for $\mathcal{E}_n$ suitable for low noise transistors (equation (6.23)), and in fact $\mathcal{E}_n$ does not start to rise until approximately f_T, at which frequency our simple model is inadequate.

The dependence of $\mathcal{E}_n$ and $\mathcal{I}_n$ on collector current and frequency for a typical bipolar transistor is shown in figure 6.10.

In the mid-frequency range, $\mathcal{I}_n$, dominated by the base current shot noise, increases with collector current. However, $\mathcal{E}_n$, dominated by collector shot noise at low collector currents, decreases as the collector current increases. Although the collector current shot noise increases with I_C, a transconductance-dependent factor g_m^{-2} (proportional to I_C^{-2}), which appears as a result of referencing the collector shot noise to the input, has a greater effect. This decrease continues until thermal noise in $r_{bb'}$ becomes dominant and $\mathcal{E}_n$ remains constant for further collector current increase. This means that at low collector currents, because F_{min} is approximately constant for transistors with low $r_{bb'}$, the inverse dependencies of $\mathcal{E}_n$ and $\mathcal{I}_n$ on I_C may be used to select R_{so} by varying I_C (see equation (6.40) and figure 6.9).

As we have seen, the frequency dependence of $\mathcal{I}_n$ at both high and low frequencies is more marked than that of $\mathcal{E}_n$ with higher flicker noise, and lower high-frequency corner frequencies.

The flicker noise $\mathcal{I}_n$ corner frequency has a slight dependence on collector current as a result of the non-linear dependence of flicker noise power on

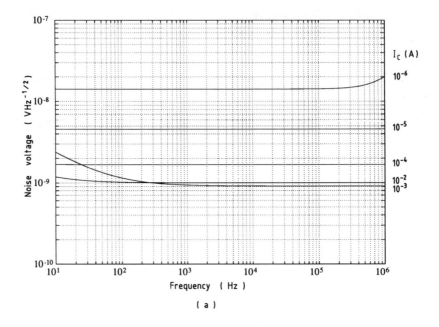

(a)

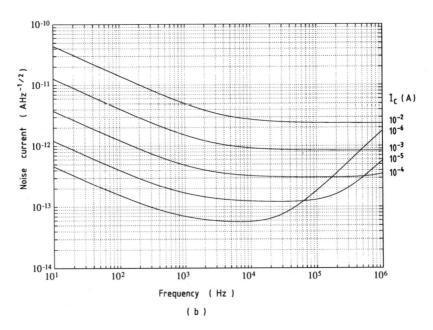

(b)

Figure 6.10 Typical variations of (a) $\mathcal{E}_n$ and (b) $\mathcal{I}_n$ with frequency and collector current for a bipolar transistor

base current ($\gamma > 1$). The high corner frequency increases more strongly with I_C as a result of its dependence on g_m (equation (6.48)).

The more rapid change of $\mathcal{I}_n$ than $\mathcal{E}_n$ at low and high frequencies means that not only does the minimum noise figure (dependent on $\mathcal{E}_n\mathcal{I}_n$ – equation (5.43)) increase, but the optimum sources resistance (dependent on $\mathcal{E}_n/\mathcal{I}_n$ – equation (5.42)) decreases as the frequency increases above the $\mathcal{I}_n$ high corner frequency and decreases below the $\mathcal{I}_n$ low corner frequency.

The noise figure dependency on collector current, in addition to source resistance and frequency, is often displayed by means of noise figure contours against source resistance and collector current at spot frequencies – as shown in figure 6.11.

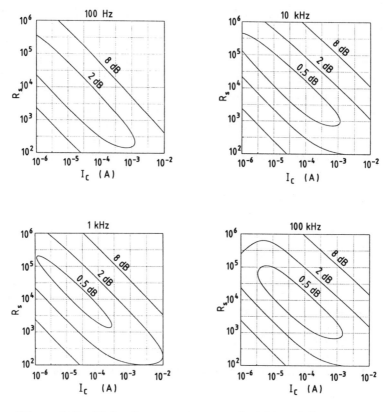

Figure 6.11   Noise figure contours against R_s and I_c

Although the above analysis has been carried out for the common emitter configuration, a similar analysis shows that the noise figure is essentially the same for the common base connection.

6.9   FET

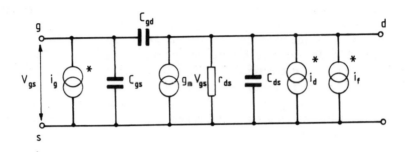

Figure 6.12   Equivalent circuit for FET

The equivalent circuit for the FET is shown in figure 6.12. Thermal noise arising from the channel resistance is shown as a noise modulation (i_d) of the drain current I_D. Since the circuit is a small signal AC model and the channel is operating under saturation conditions, the resistance associated with the thermal noise does not appear specifically. The quantity r_{ds} represents the small effect of the drain–source voltage on I_D under saturation conditions and is very high. This 'thermal' resistance of the channel (R_{ch}) is, however, related to the transconductance g_m and it is found that:

$$R_{ch} = (K_d g_m)^{-1} \tag{6.49}$$

where K_d varies very little within the normal range of operating conditions and is approximately 0.67 (van der Ziel, 1962). Thus:

$$i_D^2 \simeq 4kT(0.67)g_m \Delta f \tag{6.50}$$

The flicker noise we can write again as:

$$i_F^2 = K_F I_D^y \Delta f f^{-\alpha} \tag{6.51}$$

The sum of i_d and i_f can be referred to the input as equivalent series voltage generator e_n. Thus:

$$e_n g_m = i_d + i_f \tag{6.52}$$

and by taking mean square values, noting that i_d and i_f are independent, we obtain:

$$e_N^2 = (0.67)4kT\Delta f g_m^{-1} + K_F I_D^y \Delta f g_m^{-2} f^{-\alpha} \tag{6.53}$$

or:

$$\mathcal{E}_n = \left((0.67)4kTg_m^{-1} + K_F I_D^{\prime\prime} g_m^{-2} f^{-\alpha}\right)^{1/2} \tag{6.54}$$

This is the equivalent series voltage input noise generator.

The input parallel current generator is simply i_g, and this noise generator has two components:

$$i_N^2 = 2eI_G\Delta f + 4kT(2\pi f C_{gs})^2 g_m^{-1} K_1 \Delta f \tag{6.55}$$

or:

$$\mathcal{I}_n = \left(2eI_G + 4kT(2\pi f C_{gs})^2 g_m^{-1} K_1\right)^{1/2} \tag{6.56}$$

The first term is simply shot noise in the gate current I_G (van der Ziel, 1962). The second term is a modulation (via C_{gs}) of the gate current by the thermal noise in the channel (van der Ziel, 1963). The factor $4kT/g_m$ is related to the channel thermal noise, $2\pi f C_{gs}$ is the susceptance of C_{gs} and K_1 is a factor which is approximately constant under the normal range of operating conditions and is approximately equal to 0.25 for JFETs (van der Ziel, 1963; Robinson, 1969) and 0.1 for MOSFETs (Ambrozy, 1983).

Since channel thermal noise contributes to both e_n and i_n, these will be partially correlated. The correlation coefficient is normally small, however, and will be neglected for the rest of the analysis (Klaasen, 1967).

At frequencies sufficiently high that the flicker noise can be neglected, we have from (6.53), (6.55), (5.22) and (5.23):

$$R_{nv} = 0.67g_m^{-1} \tag{6.57}$$

$$R_{ni} = 2kT(eI_G)^{-1} \qquad \text{at low frequencies} \tag{6.58}$$

$$R_{ni} = g_m K_1^{-1}(2\pi f C_{gs})^{-2} \qquad \text{at higher frequencies} \tag{6.59}$$

Note that at very low I_G (e.g. in MOSFETs) this latter term will dominate down to very low frequencies.

From these equations and (5.38) and (5.39) we can calculate R_{so} and F_{min}. At high frequencies:

$$F_{min} = 1 + 0.8(2\pi f C_{gs})g_m^{-1} \qquad \text{for JFETs} \tag{6.60}$$

$$= 1 + 0.5(2\pi f C_{gs})g_m^{-1} \qquad \text{for MOSFETs} \tag{6.61}$$

$$R_{so} = 1.6(2\pi f C_{gs})^{-1} \qquad \text{for JFETs} \qquad (6.62)$$

$$= 2.6(2\pi f C_{gs})^{-1} \qquad \text{for MOSFETs} \qquad (6.63)$$

At low frequencies, when the gate leakage shot noise dominates $\mathcal{I}_n$, from (6.58) we have:

$$F_{min} = 1 + 2(0.33eI_G/(g_m kT))^{1/2} \qquad (6.64)$$

$$\simeq 1 + 7.3(I_G/g_m)^{1/2}$$

and:

$$R_{so} \simeq 0.2(g_m I_G)^{-1/2} \qquad (6.65)$$

Note that:

$$g_m \propto I_D^{1/2} \qquad (6.66)$$

and this indicates the use of a relatively high I_D to achieve a low F_{min}. However, the temperature rise at high I_D leads to an increase in leakage current, and since the dependence on I_D is slight this current is normally set by other design considerations.

Note also, that the $\mathcal{E}_n$ flicker noise corner frequency is often sufficiently high for flicker noise to dominate at the frequencies of interest, and it may even overlap with the high frequency component of the equivalent input current noise. This will increase both the noise factor and the optimum source resistance. In addition, $\mathcal{E}_n$ in very low noise FETs may be higher than calculated as a result of thermal noise in the parasitic resistances in series with the FET source and gate (Netzer, 1981; Brookes, 1986). For example, an FET with a g_m of 15 mS has a calculated (from (6.54)) input noise voltage of 0.85 nV Hz$^{-1/2}$ at frequencies significantly above the flicker noise corner. This corresponds to an equivalent noise resistance of 45 Ω (equation (6.57)), and a parasitic resistance comparable or greater than this will noticeably increase this noise voltage.

Typical variations of $\mathcal{E}_n$ and $\mathcal{I}_n$ with frequency are shown in figure 6.13. Some FETs exhibit an undulating low frequency noise spectrum (shown as a dashed line in the figure). This results from the random occupancy of charge-carrier traps with a range of time constants. In general, this low frequency spectrum shape changes with operating conditions. The change in $\mathcal{I}_n$ with drain current is usually small and $\mathcal{I}_n$ is often shown at only one value of I_D.

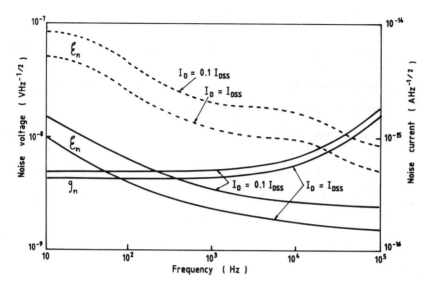

Figure 6.13 Typical variation of $\mathcal{E}_n$ and $\mathcal{I}_n$ with frequency and drain current for a FET. The dashed lines show the variation of $\mathcal{E}_n$ in a device exhibiting an undulating low frequency noise spectrum

By comparison with the bipolar transistor, the FET has a much lower noise current and comparable or greater noise voltage – leading to larger optimum source resistances – and flicker noise appears in the voltage and not the current noise generator. The low frequency noise current in MOSFETs is negligibly low, but the flicker noise corner frequency tends to be higher.

References

Ambrozy, A. (1983). *Electronic Noise*, McGraw-Hill, New York.

Baxandall, P. J. (1968). 'Noise in transistor circuits, Part 2', *Wireless World*, **74**, 454–459.

Brookes, T. M. (1986). 'The noise properties of high electron mobility transistors', *IEEE Transactions on Electron Devices*, **ED-33(1)**, 52–57.

Buckingham, M. J. (1983). *Noise in Electronic Devices and Systems*, Ellis Horwood, Chichester.

Collin, R. E. and Zucker, F. J. (1969). *Antenna Theory. Volumes 1 and 2*, McGraw-Hill, New York.

Jaeger, R. C. and Brodersen, A. J. (1970). 'Low frequency noise sources in bipolar junction transistors', *IEEE Transactions on Electron Devices*, **ED-17(2)**, 128–134.

Klaasen, F. M. (1967). 'High-frequency noise of the junction field effect transistor', *IEEE Transactions on Electron Devices*, **ED-14(7)**, 368–373.

Motchenbacher, C.D. and Fitchen, F.C. (1973). *Low-Noise Electronic Design*, Wiley, New York.

Netzer, Y. (1981). 'The design of low-noise amplifiers', *Proceedings of the IEEE*, 69(6), 728–742.

Nielsen, E.G. (1957). 'Behaviour of noise figure in junction transistor', *Proceedings of the IRE*, 45, 957–963.

Robinson, F.N.H. (1969). 'Noise in field-effect transistors at moderately high frequencies', *Electronic Engineering*, 41, 353–355.

Rotkiewicz, W. (1982). *Electromagnetic Compatability in Radio Engineering*, Elsevier, Amsterdam.

van der Ziel, A. (1962). 'Thermal noise in field effect transistors', *Proceedings of the IRE*, 50(8), 1808–1812.

van der Ziel, A. (1963). 'Gate noise in field effect transistors at moderately high frequencies', *Proceedings of the IEEE*, 51(3), 461–467.

van der Ziel, A. and Chenette, E.R. (1978). 'Noise in solid state devices', *Advances in Electronics and Electron Physics*, 46, 313–383.

7   Noise Performance Measurement

7.1 Introduction

In order to make use of analysis and low noise design techniques, the circuit designer requires a knowledge of, and/or a means of measuring, the noise performance of circuit elements, circuits and complete systems in order to determine whether or not they meet specification.

Noise measurement techniques are used by the manufacturers of semiconductors, integrated circuits and resistors in order to provide performance information to the designer. The designer may use the same techniques to check performance of the supplied devices and even to select those samples of a particular type having especially low noise; there are wide variations in noise performance, particularly $1/f$ noise generation, between samples of many circuit elements.

In this chapter we discuss the various methods of noise performance measurement, the apparatus required, the approximations and short-cuts which can sometimes be taken and, in particular, the precautions required for valid results.

7.2 Noise level measurement

7.2.1 Accuracy requirement

It is important to decide on the required accuracy at an early stage when planning noise measurements. This will determine the complexity and cost of the equipment and the effort needed. These can increase quite alarmingly with increases in required accuracy and it makes sense to measure only to an accuracy appropriate to the design problem under consideration. It is rare for the designer to need to know the rms noise level (voltage or current) to better than 10 per cent accuracy and often a knowledge within a factor of 2 is adequate.

It should be remembered that, as power is proportional to rms current or voltage squared, the uncertainty in power is greater if it is calculated from measurements of these quantities. If the fractional error in the measurement of a noise rms voltage e_N is δ_e, and that in the normalised power calculated from this measurement is δ_p then:

$$e_N^2(1 + \delta_p) = (e_N(1 + \delta_e))^2$$

or:

$$\delta_p = 2\delta_e + \delta_e^2 \qquad (7.1)$$

This equation is also true for power calculated from rms current, and actual rather than normalised power. Note that the error is approximately doubled for small δ_e.

7.2.2 Oscilloscope

The oscilloscope provides a simple – and cheap, in the sense that specialised equipment need not be purchased – method of noise level measurement if the accuracy requirements are not high. Noting that a Gaussian noise voltage is within $\pm 3\times$ rms for 99.7 per cent of the time, the rms noise voltage may be estimated to within about 20 per cent accuracy by dividing by 6 the voltage range on the screen which contains nearly all of the signal (see figure 7.1).

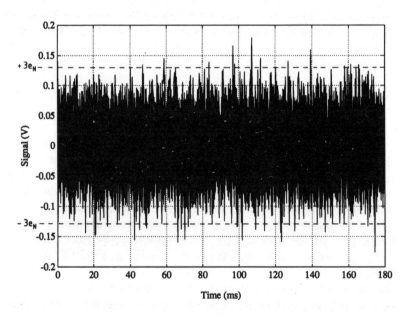

Figure 7.1   Oscilloscope display of Gaussian noise; rms level equals e_N

The oscilloscope has the advantage over a meter of easier identification of noise sources. This can lead to its use as an additional display, even when a meter or spectrum analyser is used in the interests of measurement accuracy. For example, if the noise under investigation has a significant component of popcorn noise or impulse interference, then this can be separately identified (figure 7.2). Care should be taken not to interpret the occasional high excursion of Gaussian noise as a separate noise impulse. Periodic interference – for example, at power-line frequency or from the clock, or at a subdivision of the clock frequency, of a digital circuit – is particularly easy to identify. A noise signal consisting of Gaussian noise and 50 Hz line signal is shown in figure 7.3(a).

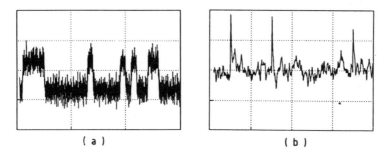

(a)   (b)

Figure 7.2   Gaussian noise plus: (a) popcorn noise; (b) impulse noise

Note that the rms random noise voltage may be estimated at the peaks of the sine wave as described above, and the rms level of the periodic signal may be measured as shown. This approach, of course, requires the time base to be synchronised with the periodic signal. The corresponding display with a free-running time base is shown in figure 7.3(b). If the periodic component is of sufficiently high amplitude compared with the random noise, the time base may be triggered from the internal signal, perhaps using a trigger filtering option, if available. At low levels of the periodic signal, external

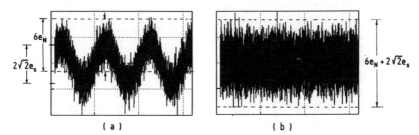

(a)   (b)

Figure 7.3   Gaussian noise (e_N rms) plus 50 Hz sine wave (e_s rms): (a) time-base synchronised to 50 Hz signal; (b) time-base free-running

triggering using the potential source of interference should be used. This technique is particularly useful when trying to identify significant noise sources and the effect of different methods of interference reduction.

Signal-to-noise ratio may be measured using a similar technique to that shown in figure 7.3(a), where the periodic signal in this case is the desired signal rather than periodic noise.

7.2.3 Audio analysis

When the noise under investigation is in the audio frequency range, it is often useful to have the option of listening to it. The ear–brain combination is a surprisingly good and easily trained audio-signal analyser and can often identify different types of noise in a mixed noise signal. For example, pure Gaussian noise has a smooth rushing sound, individual impulses can be identified as 'click' or 'plop' sounds, and the frequent impulses from a brush motor are heard as a harsh whine.

7.2.4 Meter measurements

Measurement system

When it is important to measure noise acurately then a meter is required. The meter consists of an amplifier, a rectifier, an averager and a DC meter as shown in figure 7.4.

Figure 7.4   Noise meter

Amplifier and measurement bandwidth

The amplifier should have sufficient gain to amplify the noise signal to a level suitable for the rectifier and DC meter used. Its own noise level should be small compared with the noise to be measured, although it is possible to allow for this measurement system noise, as will be shown later. The bandwidth of the amplifier should be greater than that of the noise to be measured when making a broadband noise measurement. It is important to note that it may need to be very significantly greater, depending on the shape of the noise spectrum and that of the amplifier frequency response. For example, if the noise is white noise, low-pass filtered by a single pole CR

network, and the amplifier is similarly high frequency limited by a single pole filter, then the spectrum of the noise at the input of the amplifier is:

$$S_n(f) = \frac{S_n(0)}{1 + (f/f_n)^2} \tag{7.2}$$

and at the output:

$$S_{on}(f) = S_n(f)G_a(f) \tag{7.3}$$

where:

$$G_a(f) = G_a(0)(1 + (f/f_a)^2)^{-1} \tag{7.4}$$

is the frequency-dependent, normalised power gain of the amplifier, where f_n and f_a are the 3 dB cut-off frequencies of the input noise and amplifier response respectively and $S_n(0)$ is the spectral density of the input noise at low frequencies ($f \ll f_n$). The spectra are shown in figure 7.5.

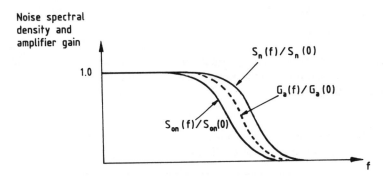

Figure 7.5   Spectral densities of noise at input and output of amplifier, together with the normalised power gain frequency response – all normalised to unity at $f = 0$

The actual and measured normalised noise powers are, respectively:

$$e_N^2 = \int_0^\infty S_n(f)df$$

$$= \frac{\pi}{2} S_n(0)f_n \tag{7.5}$$

and:

$$e_{On}^2 = \int_0^\infty S_{on}(f)\mathrm{d}f/G_a(0)$$

$$= \frac{\pi}{2(1 + 1/\beta)} S_n(0) f_n \tag{7.6}$$

where $\beta = f_a/f_n$, and we have made use of the results:

$$\int_0^\infty \frac{\mathrm{d}x}{(1 + x^2)} = \pi/2$$

$$\int_0^\infty \frac{\mathrm{d}x}{(1 + x^2)(1 + (x/\beta)^2)} = \frac{\pi}{2(1 + 1/\beta)} \tag{7.7}$$

The fractional measurement error is:

$$\varepsilon_a = \frac{e_{On}^2 - e_N^2}{e_N^2}$$

$$= -(1 + \beta)^{-1} \tag{7.8}$$

and is shown for a range of β values in table 7.1. It is clear that a measurement bandwidth very much greater than the noise bandwidth is required for accurate noise power measurement if the bandwidths are limited in this way.

Table 7.1 Error in measured normalised noise power against ratios of amplifier-to-noise bandwidth.

Ratio of measurement to noise bandwidths β	Measurement error ε_a (%)
1	−50.0
2	−33.3
5	−16.7
10	−9.1
20	−4.8
50	−2.0
100	−1.0

When making spot frequency measurements, the amplifier includes a narrow band filter, although if more than a few spot frequency measurements are required it is easier to use a spectrum analyser to display the complete noise spectrum over the frequency range of interest. The width of this filter is important. The term 'spot frequency' implies that the noise spectral density is measured at one frequency. In practice, we have to measure the noise power within a narrow frequency band Δf in order to calculate spectral density and, as will be shown later in this section, the time required to complete a measurement to a given accuracy is inversely proportional to this bandwidth. With a narrow bandwidth of a few Hertz and a reasonable accuracy requirement, this time can amount to several minutes. There is, therefore, good reason to use as wide a measurement bandwidth as possible.

The deterministic error in spectral density measurement is determined by the change in spectral density of the noise over the bandwidth of the measurement filter. Considering the measurement of spectral density at frequency f_0 using a filter and amplifier combination having response $G_m(f)$ with centre frequency f_0, then the measured mean square voltage is:

$$e_M^2 = \int_0^\infty S_x(f) G_m(f) \, df \tag{7.9}$$

where $S_x(f)$ is the spectral density of the input noise. If $G_m(f)$ is sufficiently narrow that $S_x(f)$ can be considered constant and equal to $S_x(f_0)$ over the extent of $G_m(f)$ then:

$$e_M^2 = S_x(f_0) \int_0^\infty G_m(f) \, df \tag{7.10}$$

$$= S_x(f_0) G_{po} B_m \tag{7.11}$$

where G_{po} is the gain at frequency f_0 and B_m, the measurement bandwidth, is:

$$B_m = \frac{\int_0^\infty G_m(f) \, df}{G_{po}} \tag{7.12}$$

Note that this is the noise bandwidth (see section 4.2.5) of the measurement system and it is equal to the Δf used in the other chapters.

The spectral density at $f = f_0$ is (from equation (7.11)):

$$S_x(f_0) = \frac{e_M^2}{G_{po} B_m} \tag{7.13}$$

The effect of small variations of $S_x(f)$ over the measurement bandwidth can be analysed by expressing $S_x(f)$, within the extent of $G_m(f)$, as a Taylor series expansion about its value at $f = f_0$ and considering the first three terms. Then:

$$S_x(f) \simeq S_x(f_0) + S_x'(f_0)(f - f_0) + \frac{1}{2}S_x''(f_0)(f - f_0)^2 \qquad (7.14)$$

The measured mean square voltage, from equation (7.9), is now:

$$e_M^2 = S_x(f_0)G_{po}B_m + S_x'(f_0)\int_0^\infty G_m(f)(f - f_0)df$$

$$+ \frac{1}{2}S_x''(f_0)\int_0^\infty G_m(f)(f - f_0)^2 df \qquad (7.15)$$

Normally, the measurement filter response is symmetrical about f_0. In this case, the integral of $G_m(f)(f - f_0)$ is zero. Thus, if $G_m(f)$ is symmetrical, a linear variation of $S_x(f)$ will not give rise to an error in measurement. If there is significant curvature of $S_x(f)$ – that is, $S_x''(f_0)$ is not negligible then, from equation (7.15):

$$e_M^2 = S_x(f_0)G_{po}B_m + \frac{1}{2}S''(f_0)G_{po}B_m\sigma_m^2 \qquad (7.16)$$

where σ_m^2 is the mean square width of $G_m(f)$. That is:

$$\sigma_m^2 = \frac{\int_0^\infty G_m(f)(f - f_0)^2 df}{\int_0^\infty G_m(f)df} \qquad (7.17)$$

The measured spectral density at $f = f_0$ is now (from (7.16)):

$$S_{mx}(f_0) = \frac{e_M^2}{G_{po}B_m} = S_x(f_0)\left(1 + \frac{1}{2}\frac{S_x''(f_0)}{S_x(f_0)}\sigma_m^2\right)$$

and the fractional error is:

$$\varepsilon_m = \frac{1}{2}\frac{S_x''(f_0)}{S_x(f_0)}\sigma_m^2 \qquad (7.18)$$

For a rectangular $G_m(f)$, this is (making use of equation (7.17)):

$$\varepsilon_m = \frac{B_m^2}{24}\frac{S_x''(f_0)}{S_x(f_0)} \qquad (7.19)$$

For $1/f$ noise this is:

$$\varepsilon_m = \frac{B_m^2}{12f_0^2} \tag{7.20}$$

which suggests that a fairly high fractional measurement bandwidth B_m/f_0 may be used in this case. For example, if $B_m/f_0 = 0.5$, the fractional error is only about 2 per cent. The above points are illustrated in figure 7.6.

The amplifier should also be operated such that it does not limit except at the occasional extreme values of the input noise, remembering that a sinusoidal wave is contained within only $\pm\sqrt{(2)}\times$ rms, whereas there are a significant number of excursions of a Gaussian waveform up to ± 3 or $4\times$rms.

Rectifier

Preferably, the measurement system should give an indication of true rms value – giving an accurate measurement of rms voltage, whatever the input waveform shape. In this type of measurement system the amplified noise signal is passed to a squaring circuit or square law rectifier; the output of the averager is then the mean square voltage, and either the meter scale is calibrated to indicate the square root of the applied voltage (only applicable in the case of an analogue or pointer meter), or an analogue or digital square rooting circuit is used (between averager and meter in figure 7.4) before display on a linear scale meter.

Some systems use a diode rectifier to measure alternating voltages and these give a reading which depends on the input waveform and are not true rms indicating. These measurement systems will normally be calibrated for sine-wave inputs. The mean value of a full wave rectified sine wave is $(2\sqrt{(2)}/\pi)\times$ rms and this reading is multiplied by $\pi/(2\sqrt{2})$ before display. However the mean of a full wave rectified Gaussian noise signal is $\sqrt{(2/\pi)}\times$ rms, so that the meter calibrated to indicate correctly the rms level of a sine wave gives a reading of:

$$\pi/(2\sqrt{2}) \times \sqrt{(2/\pi)} \times \text{rms} = 0.886 \times \text{rms}$$

with Gaussian noise.

If such a meter is used for the measurement of Gaussian noise, the reading should be multiplied by $1/0.886 = 1.13$. It will not, of course, give the correct measurement with mixed noise. Note that the average output of a half-wave rectifier is half that of a full wave for both sine and Gaussian waveforms, so the same correction factor may be used in this case.

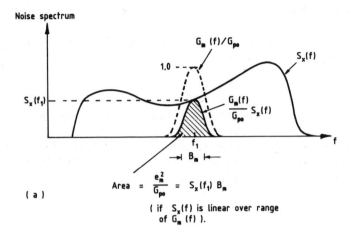

(a)

$$\text{Area} = \frac{e_m^2}{G_{po}} = S_x(f_1) B_m$$

(if $S_x(f)$ is linear over range of $G_m(f)$).

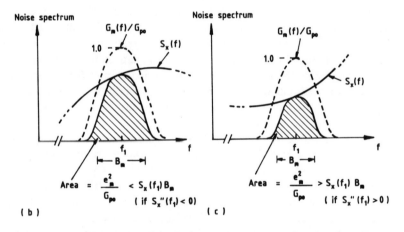

(b)

$$\text{Area} = \frac{e_m^2}{G_{po}} < S_x(f_1) B_m$$
(if $S_x''(f_1) < 0$)

(c)

$$\text{Area} = \frac{e_m^2}{G_{po}} > S_x(f_1) B_m$$
(if $S_x''(f_1) > 0$)

Figure 7.6 Spectral density measurement using narrowband filter. Effect of variation of noise spectrum over measurement passband: (a) linear variation $S''(f_1)=0$; (b) non-linear variation $S''(f_1)<0$; (c) non-linear variation $S''(f_1)>0$

Averaging

The output of squarer or diode rectifier ($y(t)$ in figure 7.4) is a random quantity and requires smoothing or averaging before display. This can be a running average, using a low-pass filter, such that:

$$z(t) = y(t) * h_{\mathrm{L}}(t) \tag{7.21}$$

where $h_{\mathrm{L}}(t)$ is the impulse response of the low-pass filter, or an integration over a period of time τ such that:

$$z(t) = \frac{1}{\tau} \int_{t-\tau}^{t} y(t)\mathrm{d}t \tag{7.22}$$

In practice, the integration is usually carried out on consecutive signal segments of duration τ, rather than continuously, leading to averages updated at intervals $n\tau$ (n = 1, 2, 3 ...). That is:

$$z(n\tau) = \frac{1}{\tau} \int_{(n-1)\tau}^{n\tau} y(t)\mathrm{d}t \tag{7.23}$$

With a moving coil meter, the low-pass filtration may be provided by mechanical damping of the meter movement.

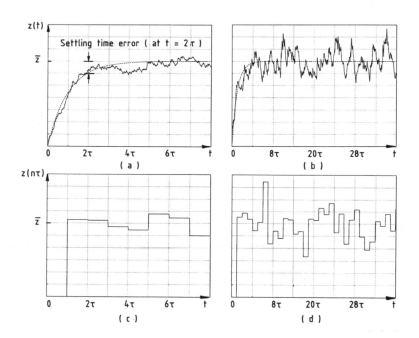

Figure 7.7   Averager output: (a, b) single pole CR low-pass filter –
(a) large τ ($\tau = CR$); (b) small τ ($\tau = CR$); (c, d) integrator –
(c) large τ; (d) small τ

The averaged squared or rectified signal has a mean value and a random variation (about this mean) which reduces as the averaging time (the time constant of the low-pass filter or the integration time) increases (figure 7.7). The degree of random variation and its relationship to the signal bandwidth and averaging time are of interest since these limit the accuracy of any single reading. In general, for Gaussian noise, the rms deviation of the reading from the mean, expressed as a fraction of the mean is given by:

$$\delta_r = \frac{\text{rms deviation of meter reading from mean}}{\text{mean meter reading}}$$

$$= \frac{1}{k(B\tau)^{1/2}} \tag{7.24}$$

where B is the bandwidth (B_n) of the noise signal when making broadband measurements, or the measurement bandwidth $(B_m = \Delta f)$ when making spot frequency measurements, and τ is the averaging time. The value of k depends on the spectrum shape, the definition of B, the type of averaging and whether a squarer or diode rectifier is used. We can write:

$$k = k_1 k_2 k_3 \tag{7.25}$$

where the values of k_1, k_2, k_3 are determined by the type of rectifier, the noise spectrum and the type of averaging respectively (Davenport and Root, 1958; Middleton, 1960; Harman, 1963; Blachman, 1966; Bracewell, 1986). Examples are given in table 7.2. For these values of k_2, the B in equation (7.24) is the 3 dB bandwidth in each case. Table 7.2 gives good approximation to k. It has been assumed that $B\tau \gg 1$ and that, in the case of the half-wave rectifier, there are no signal frequency components of the order of, or less than, $1/\tau$. The analysis, in the case of the diode rectifiers, involves neglecting small terms in a series expansion of the post-rectifier spectrum, and the product $k_1 k_2$ will be in error by a few per cent depending on input spectrum shape. The expression for k_2, in the case of filtered $1/f$ noise, is frequency dependent since the spectral shape is not constant. For most practical cases, it may be put equal to unity.

As might be expected, the values of k_1 indicate that the fractional error when measuring mean square voltage (normalised power) is twice that when measuring the corresponding root mean square voltage. The apparent slight advantage of using a running average CR filter as opposed to an integrator is illusory, because we have to wait for the output of the filter to reach a stationary condition (figure 7.7) before taking a measurement, whereas the output of the integrator is available after the integration period.

Table 7.2 Factors (k_1, k_2, k_3) determining measurement error

Type of rectifier	k_1
Square law	1
Square law (square root taken after averager – true rms reading)	2
Full or half wave diode	2

Input Spectrum	k_2
White noise filtered by: rectangular bandpass or low-pass filter	1
single CR low-pass filter	$\pi^{1/2}$
Gaussian bandpass filter	$\left(\dfrac{\pi}{2\ln 2}\right)^{1/4}$
$1/f$ noise filtered by rectangular bandpass filter with passband from f_1 to f_2	$\dfrac{(f_1 f_2)^{1/2}\ln(f_2/f_1)}{f_2 - f_1}$
	$= 1.0 \qquad f_2/f_1 = 1.1$
	$= 0.98 \qquad f_2/f_1 = 2.0$
	$= 0.81 \qquad f_2/f_1 = 10$
	$= 0.47 \qquad f_2/f_1 = 100$

Averaging	k_3
Single CR low-pass filter ($\tau = CR$)	$2^{1/2}$
Integrator	1

It is important to note that the capacitor of the running averager should have some degree of isolation from the output of the diode rectifiers, since a capacitor coupled directly to the output will create a peak-reading or envelope-following demodulator. Specifically, the charge and discharge times of the capacitor should be approximately equal. The circuit in figure 7.8 is suitable if R_2 is very much greater than R_1. In this case τ is approximately equal to CR_2. Note that if the meter resistance is not very much greater than R_2 it will reduce the time constant and attenuate the output.

It is instructive to consider the integration times required for typical situations. From equation (7.24) we have:

$$\tau = \frac{1}{k^2 \delta_r^2 B} \tag{7.26}$$

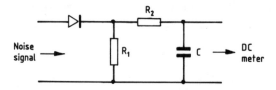

Figure 7.8   Half-wave rectifier plus averaging single pole CR low-pass filter

For the case of rectangular bandwidth noise, an rms indicating meter and a linear integrator, this is:

$$\tau = \frac{1}{4\delta_r^2 B} \tag{7.27}$$

If $B = 10\,\text{kHz}$ (wideband audio frequency noise) and we require a fractional rms error of 5 per cent ($\delta_r = 0.05$), then $\tau = 10\,\text{ms}$.

If, however, we are making a spot frequency measurement at low frequencies – measuring $1/f$ noise in low frequency amplifiers, for example – and have a measurement bandwidth of 10 Hz, then the same rms error requires an integration time of 10 s. An error of 1 per cent would require an integration time of 250 s. If a running average CR filter is used, then these times (the time constant CR) are halved, but the settling time must be taken into account. There will be an additional error, arising from the settling time, of 5 per cent after a time $3CR$, or 1.8 per cent after a time $4CR$ (figure 7.7). These figures illustrate one of the problems of making low bandwidth noise measurements.

7.2.5 Spectrum analysis

Instead of making a number of spot measurements of noise spectral density, it will often be more convenient to use a spectrum analyser to obtain a complete spectral density over the frequency range of interest. In addition to its use in measuring the spectral density of random noise, it is very useful in indicating the line spectra of periodic interference (figure 7.9) and the effectiveness of interference reduction techniques.

It is beyond the scope of this book to discuss the different methods of spectral analysis in detail: it is an extensive field and there are many good texts on the subject. See, for example, Bracewell (1986), Bendat and Piersol (1971), Lathi (1968) and Papoulis (1984). However, a brief description of the most common techniques is of value.

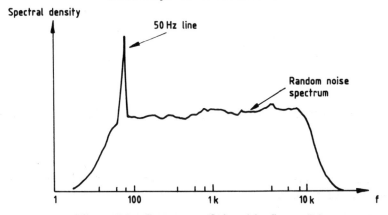

Figure 7.9   Spectrum of signal in figure 7.3

The conceptually easiest method of spectral analysis is simply to make a number of spot frequency measurements using a narrowband filter of adjustable centre frequency. This is difficult to achieve in practice – particularly at low frequencies. A more usual arrangement (shown in figure 7.10) is to mix the signal with a higher frequency sine wave from a voltage controlled oscillator (VCO), so shifting the signal spectrum to a higher frequency range, and to pass this shifted-spectrum signal to a fixed frequency, narrow bandpass filter. The output of this filter is proportional to the spectral density of the input signal at a frequency equal to the difference between the fixed filter centre frequency and the VCO frequency. The spectrum is measured by sweeping the VCO frequency such that the frequency-shifted input signal spectrum is swept past the fixed filter frequency (figure 7.11).

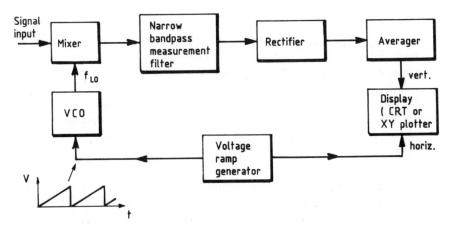

Figure 7.10   Swept-frequency spectrum analyser

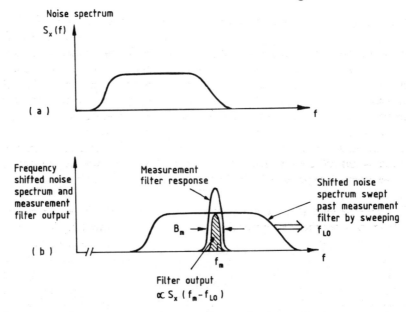

Figure 7.11 (a) Noise spectrum and (b) frequency-shifted noise spectrum and measurement filter output. Local oscillator frequency is f_{LO} (swept). Measurement filter centre frequency is f_m and bandwidth is B_m

It is important to make sure that the sweep speed is sufficiently slow to allow the post-rectifier averaging filter to respond to changes in spectral density. An experimental method to determine whether or not the rate is sufficiently slow is to make successive measurements while reducing the frequency sweep rate until the displayed spectrum no longer changes. Note that the product of measurement bandwidth B_m and averaging time constant τ should be determined by accuracy requirements as discussed in section 7.2.4. This means that τ is usually very much greater than $1/B_m$ and the settling time of the filter (of the order of $1/B_m$) will have negligible effect on the settling time of the output. When such an instrument is used for measuring the spectrum of deterministic signals, a low τB_m may be used and the filter response time may have to be taken into account when determining scan rate.

The requirement of a large τB_m product when measuring noise spectra leads to scan times of many minutes at low frequencies. An alternative is the measurement of the spectral density at a number of frequencies simultaneously using a bank of filters. This is a rather expensive and inflexible arrangement, so a much preferred method of achieving the same end is spectral estimation by digital signal processing. A favourite method makes use of the fast Fourier transform (FFT) – a rapid (as the name implies) digital method of calculating the Fourier transform of the signal. As has been explained in chapter 2, the spectrum may be calculated from the squared modulus of the Fourier transform. A high proportion of commercial spectrum analysers use this method of spectral analysis.

The necessary calculations are carried out by software using a microprocessor-controlled instrument. The core FFT process may be performed by a dedicated integrated circuit. The process is illustrated in figures 7.12 and 7.13.

The method estimates the power spectrum of the signal from the energy spectrum of a number of segments (duration τ_s) of the input signal. These segments are shown as adjoint (figure 7.13) but may be overlapping. The digitised signal in each segment is multiplied by a window function which gradually tapers off the contribution of the signal towards the beginning and end of the segment, and an FFT is performed on the windowed signal. The function of the tapering window is to reduce spurious peaks in the measured spectrum resulting from a sharp signal 'turn-on' and 'turn-off' at the beginning and end of the data segment. The estimate of the power spectrum is then calculated from the squared modulus of the Fourier transform. The random error in the estimate is high. In fact, since the spectral resolution of the FFT is approximately the reciprocal of the data length ($B_m \simeq 1/\tau_s$) and the effective averaging time is the length of the data segment (τ_s), the fractional error is:

$$\varepsilon_r \simeq \frac{1}{(B_m\tau_s)^{1/2}} = 1 \qquad (7.28)$$

That is, the rms error is approximately equal to the mean. The exact error is dependent on the window function used.

A better spectral estimate is formed from an average of the spectra from a number (N_s) of data segments, as shown in figure 7.13. Since the rms error in averaged random variables is reduced by the square root of the number averaged, the fractional error in this case is:

$$\varepsilon_r \simeq (B_m N_s \tau_s)^{-1/2} = N_s^{-1/2} \qquad (7.29)$$

If the signal segments overlap then the error is larger than this, as a result of partial correlation of the signals in overlapping segments.

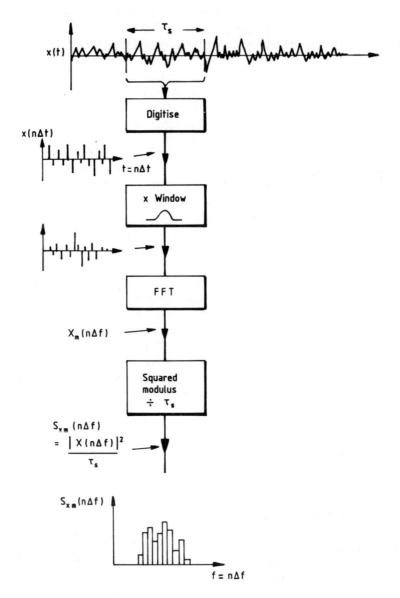

Figure 7.12 Process of estimating the noise spectrum from one signal segment using the fast Fourier transform (FFT). The time sampling interval is Δt. The estimates $X_m(n\Delta f)$ and $S_{xm}(n\Delta f)$ of the Fourier transform and spectrum respectively of $x(t)$ are calculated at frequency intervals $\Delta f = 1/\tau_s$

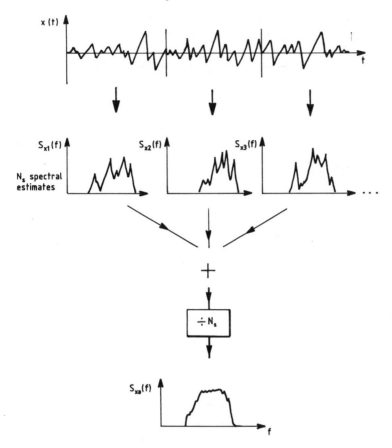

Figure 7.13 Averaging of spectral estimates from N_s signal segments to give improved estimate $S_{xa}(f)$

7.3 Noise figure measurement

7.3.1 Introduction

The noise figure, as we have seen in chapter 5, section 5.2.9, is a figure of merit of two-port networks which indicates the degree to which the network reduces the signal-to-noise ratio of any signal source by adding its own noise to the unavoidable thermal noise of the source. In order to make choices between devices or systems, we need simple methods of measuring their noise figures. Because we can easily calculate the thermal noise of a source from its impedance, the core of the calculation of the noise figure is the measurement of the equivalent noise at the input of the network. This can be achieved by making two measurements – comparing the response of the network to the equivalent input noise and to a calibrated signal generator using the arrangement shown in figure 7.14. This signal generator can be either a sine-wave or a broadband noise generator (Haus *et al*, 1960).

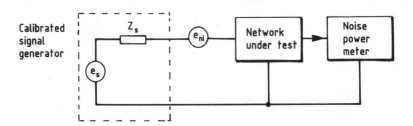

Figure 7.14 General layout for noise-figure measurement. The equivalent input noise e_{ni} arises from the network equivalent input noise generators and thermal noise in Z_s

7.3.2 Sine-wave method

For a spot frequency measurement, the sine-wave signal generator is tuned to the centre frequency (f_0) of the filter in the power meter (strictly, the frequency at which the gain G_{po} of the meter filter is measured in order to calculate its noise bandwidth – see chapter 4, section 4.2.5). For the first measurement (noise only), the generator output is set at zero and the measured normalised noise power output (e_{ON}^2) of the network within the measurement bandwidth is noted. The output signal of the generator is then increased, for the second measurement, to give a measured normalised output (noise plus sine-wave) from the amplifier of e_O^2. The normalised output power arising from the generator input alone is then:

$$e_{OS}^2 = e_O^2 - e_{ON}^2 \qquad (7.30)$$

If the open-circuit output of the signal generator is e_S V rms, then the normalised power gain, from signal generator to network output, is:

$$G(f_0) = e_{OS}^2 / e_S^2 \qquad (7.31)$$

Note that this gain includes the attenuator formed by the source impedance Z_s and the input impedance of the amplifier, and is different from both the available power gain and the normalised power gain (network voltage gain squared).

But we also have:

$$e_{ON}^2 = G(f_0)e_{Ni}^2 \tag{7.32}$$

where e_{Ni} is the total equivalent input noise voltage (see chapter 5, section 5.2.6).

Note that both e_{ON}^2 and e_{Ni}^2 are the normalised noise powers only within the measurement bandwidth.

Eliminating the gain from the previous two equations gives:

$$e_{Ni}^2 = \frac{e_S^2}{e_{OS}^2}e_{ON}^2 \tag{7.33}$$

If the series resistance component of Z_s is R_s, the source normalised thermal noise power is $4kTR_sB_m$ where B_m is the measurement noise bandwidth, and the noise factor is (see chapter 5, section 5.2.9):

$$F = \frac{e_{Ni}^2}{4kTR_sB_m} \tag{7.34}$$

or, making use of equations (7.30) and (7.33):

$$F = \frac{e_S^2}{4kTR_sB_m}\frac{1}{\dfrac{e_O^2}{e_{ON}^2} - 1} \tag{7.35}$$

This can be simplified in one of two ways. If, for the second measurement, the signal generator output is set such that the indicated output power is double that of the first measurement, then the signal generator output must be equal to the equivalent input noise and:

$$F = \frac{e_S^2}{4kTR_sB_m} \tag{7.36}$$

Since half the output signal is noise during the second measurement, the required averaging times are comparable with those required for noise only.

Alternatively, if, for the second measurement, the signal generator output is increased until the sine-wave signal output from the amplifier is much greater than the noise output, then:

$$F = \frac{e_S^2}{4kTR_sB_m}\frac{e_{ON}^2}{e_O^2} \tag{7.37}$$

This may be written:

$$F = \frac{e_{ON}^2}{4kTR_sG(f_0)B_m} \tag{7.38}$$

since, in this case:

$$G(f_0) = \frac{e_O^2}{e_S^2} \tag{7.39}$$

The output power doubling method has the advantage of simplicity but is not easy to implement with narrow measurement bandwidths because the long integration times required make adjustment of the signal generator output to give a particular output power reading difficult. The high-level sine-wave method has the advantage that only one noise measurement is required and the integration time may be reduced for the sine-wave measurement. However, it is not always possible to achieve the condition $e_{OS}^2 \gg e_{ON}^2$ without driving the device or system under test into non-linearity.

Particularly at high radio frequencies, the networks under test are often impedance matched to the source and to the power meter, and it is convenient to deal with available, rather than normalised, power and power gain. Noting that the available power from the signal generator is:

$$P_s = \frac{e_S^2}{4R_s} \tag{7.40}$$

and writing the available power outputs with noise, noise plus signal and signal only inputs as P_{on}, P_o and P_{os} respectively, the equations corresponding to (7.35)–(7.39) are:

in general

$$F = \frac{P_s}{kTB_m} \frac{1}{\dfrac{P_o}{P_{on}} - 1} \tag{7.41}$$

for the power doubling method

$$F = \frac{P_s}{kTB_m} \tag{7.42}$$

and for the high-level sine-wave method

$$F = \frac{P_s P_{on}}{kTB_m P_o} \tag{7.43}$$

or

$$F = \frac{P_{on}}{kTG_a(f_0)B_m} \tag{7.44}$$

where $G_a(f_0)$ is the available gain of the network under test at frequency f_0.

The broadband measurements are similar except that the noise bandwidth is now that of the network under test. The relevant equations may be derived by substituting the noise bandwidth B_n of the network for B_m and mid-band gain G_{po} for $G(f_0)$ in equations (7.35)–(7.44) (mid-band available gain in (7.44)).

Note that, for the broadband measurement, the method requires a knowledge of the network noise-bandwidth, and the sine-wave signal frequency (f_0) should be that at which G_{po} (or the available gain) is measured, although this condition may be relaxed as long as the passband gain at this frequency is not significantly different from that at f_0.

If the noise bandwidth is not known, then it may be calculated using equation (4.17). The power gain $G_p(f)$ may be measured from the variation in output power as the signal frequency is altered. Note that this may be normalised power gain e_{os}^2/e_s^2 or available power gain P_{os}/P_s depending on whether the broad-band equivalents of the equations (7.35)–(7.38) or (7.41)–(7.44) are being used. If it is not possible to use sufficiently high signal levels such that the noise output is negligible by comparison, then the power output versus frequency will appear as shown in figure 7.15. The integral should be taken using the noise output as the base line, not the zero power level (Haus *et al.*, 1960).

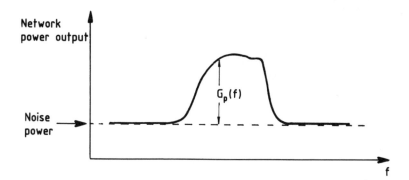

Figure 7.15   Measurement of power gain frequency response

7.3.3 Noise generator method

In the noise generator method, the signal source is a broadband noise source with uniform and known spectral density over the passband of interest. Like the sine-wave method, this method may be used to measure spot frequency or average noise factor using normalised or available powers and power gains. As in the sine-wave method, the output power of the network under test is measured firstly with the noise generator output set at zero (usually switched off) and secondly with a non-zero output. In fact, the same equations may be used. The difference is that the output of the noise generator is specified in terms of its (constant) spectral density. For narrow-band, normalised power measurements this is:

$$P' = \frac{e_S^2}{B_m} \tag{7.45}$$

for average (wideband) measurements:

$$P' = \frac{e_S^2}{B_n} \tag{7.46}$$

and for measurements expressed in terms of available power, the available power spectral densities of the noise generators are:

$$P'_a = \frac{P_s}{B_m}$$

and: $\tag{7.47}$

$$P'_a = \frac{P_s}{B_n}$$

for spot frequency and average noise factor measurements respectively.

Making use of the above in equations (7.35)–(7.37) and (7.41)–(7.43), we have:

For the general case

$$F = \frac{P'}{4kTR_s} \frac{1}{\dfrac{e_O^2}{e_{ON}^2} - 1} \tag{7.48}$$

or

$$F = \frac{P'_a}{kT} \frac{1}{\dfrac{P_o}{P_{on}} - 1} \qquad (7.49)$$

For the power doubling method

$$F = \frac{P'}{4kTR_s} \qquad (7.50)$$

or

$$F = \frac{P'_a}{kT} \qquad (7.51)$$

and for the high signal input method

$$F = \frac{P'}{4kTR_s} \frac{e^2_{ON}}{e^2_O} \qquad (7.52)$$

or

$$F = \frac{P'_a}{kT} \frac{P_{on}}{P_o} \qquad (7.53)$$

These formulae are valid for both the average and spot frequency measurements. The important advantage of this method is that a prior knowledge or measurement of the noise bandwidth is not required.

There are a number of variations on this method. One makes use of the arrangement shown in figure 7.16. The first output power measurement, with the noise generator output set to zero, is made with the output signal attenuator set to give a suitable (high scale, for good resolution) power meter reading. The attenuation is then increased by 3 dB and the noise generator output increased until the same power meter reading is obtained. Under these conditions, the output noise power from the network has doubled and we may use equation (7.50) or (7.51). The advantage of this variant is that an accurate attenuator rather than the more expensive accurate power meter is required.

An interesting variation involves using thermal noise – a resistor as the noise source (Baxandall, 1968). The method is illustrated in figure 7.17. The output power is measured with the same resistor at the temperature T at which the noise figure is required, and at a temperature T_1 at which the

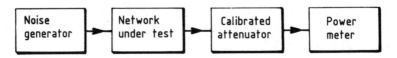

Figure 7.16   Output noise power doubling method of noise figure
measurement using a calibrated attenuator

thermal noise is significantly different. Immersion in liquid nitrogen is a
suitable means of achieving this. Note that it is important that the resistance
remains substantially unchanged between these two temperatures. An
alternative is switching between two similar resistances at these two
different temperatures.

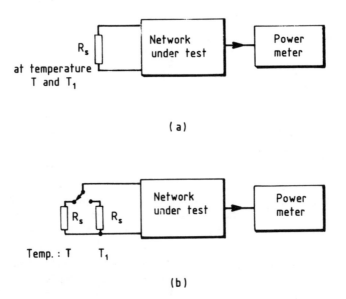

Figure 7.17   Noise figure measurement using source resistance R_s at two
temperatures: (a) single resistor; (b) two resistors

The normalised output powers for source temperatures T and T_1 are
respectively:

$$e_{o1}^2 = (4kTR_sB_m + e_A^2)G_{po} \qquad (7.54)$$

$$e_{o2}^2 = (4kT_1R_sB_m + e_A^2)G_{po} \qquad (7.55)$$

where e_A^2 is the amplifier noise contribution referred to the input. For example, with uncorrelated equivalent input noise generators and resistive source (equations (5.21)):

$$e_A^2 = e_N^2 + i_N^2 R_s^2 \tag{7.56}$$

but equations (7.54) and (7.55) are valid in the general case of correlated generators and complex source impedance.

In addition, by definition:

$$F = \frac{e_{o1}^2}{4kTR_s B_m G_{po}} \tag{7.57}$$

Subtracting equation (7.55) from (7.54), dividing by e_{o1}^2 derived from equation (7.57) and rearranging gives:

$$F = \frac{1}{1 - \dfrac{e_{o2}^2}{e_{o1}^2}} \, \frac{T - T_1}{T} \tag{7.58}$$

Note that the same equation may be used for wideband (average) noise factor calculation and the output ratio (e_{o2}^2/e_{o1}^2) may be replaced by the available output power ratio.

If T_1 is greater than T, it is more convenient to write equation (7.58) as:

$$F = \frac{1}{\dfrac{e_{o2}^2}{e_{o1}^2} - 1} \, \frac{T_1 - T}{T} \tag{7.59}$$

This method is sometimes called the Y-factor method (Adler *et al.*, 1963). The ratio of 'hot' to 'cold' source available output powers is:

$$\frac{e_{o2}^2}{e_{o1}^2} = \frac{P_{o2}}{P_{o1}} = Y \tag{7.60}$$

One means of obtaining a 'hot' source without physically raising its temperature is to have a noise generator with the same source impedance and an equivalent temperature of T_1. To demonstrate this, we can alter equation (7.48) to express F in terms of the equivalent noise temperature (T_{ne}) of the noise generator. The thermal noise from R_s raised to temperature T_{ne} must equal the noise generator output (including thermal noise in R_s). Thus:

$$4kT_{ne}R_s = P' + 4kTR_s \tag{7.61}$$

Eliminating P' from the above and equation (7.48), and identifying T_1 with T_{ne}, gives equation (7.59).

7.3.4 Temperature correction

Since thermal source-noise, and therefore noise factor, are temperature dependent, it is useful to have a standard noise temperature for comparison purposes. This temperature is 290 K, or 17°C to the nearest degree.

If the noise factor has been measured at another source temperature, then a correction should be applied to obtain the standard temperature noise factor.

The measured and standard noise factors are respectively:

$$F = 1 + \frac{e_A^2}{4kTR_sB_m} \tag{7.62}$$

and:

$$F_{std} = 1 + \frac{e_A^2}{4k(290)R_sB_m} \tag{7.63}$$

Eliminating e_A^2 and rearranging gives (Haus *et al.*, 1960):

$$F_{std} = 1 + (F-1)T/290 \tag{7.64}$$

The correction is usually small. For example, if T is 310 K (37°C) and F is 2 (3 dB), then F_{std} is 2.07 (3.16 dB).

7.3.5 Mixing down

For noise measurements at very high radio frequencies, it is often convenient to translate the noise spectrum to be measured to a lower frequency using the arrangement shown in figure 7.18 (Hewlett Packard, 1983). Any component of the noise output from the network under test at frequency f will contribute to the measurement if $|f-f_{LO}|$ is within the passband of the intermediate freqency (IF) amplifier (bandwidth B_m centred at f_{IF}) (Hewlett Packard, 1988). This means that noise from the network-under-test in two frequency bands (sidebands) of width B_m and centred at $f_{LO}-f_{IF}$ and $f_{LO}+f_{IF}$ (as shown in figure 7.19) contribute to the measurement, and the noise spectral density assigned to frequency f_{LO} is the average of the noise spectrum over these two bands. This can lead to an error, as shown, if there is a significant variation in noise spectral density between these two bands. The network under test may also see a significantly different source impedance (if frequency dependent) within these two bands. In addition, if the local oscillator signal contains significant harmonics at nf_{LO}, then noise in frequency bands around $nf_{LO} \pm f_{IF}$ will contribute to the measured power and lead to error. These problems can be alleviated by the use of a pre-mixer filter between the network and mixer with a frequency centred on either $f_{LO} + f_{IF}$ or $f_{LO} - f_{IF}$, and with bandwidth greater than B_m but sufficiently narrow to reduce noise from the unwanted sideband and harmonic responses to negligible levels. The tuning of this filter must be linked to that of the local oscillator.

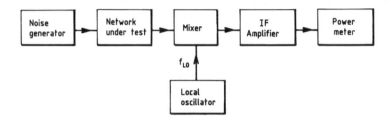

Figure 7.18 Arrangement for the measurement for high radio-frequency noise using a down-mixer

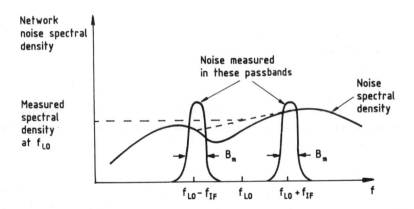

Figure 7.19 Spectral density measurement error arising from double sideband response

Care should be taken when a mixer or complete heterodyne system is the device under test. Normally a mixer is used in a system responding only to signals with frequencies in a sideband centred on either $f_{LO} - f_{IF}$ or $f_{LO} + f_{IF}$, and signals with frequencies in the other band (image frequencies) are filtered out before the mixer. In this case, the source thermal noise in the denominator of the noise factor equation should be that within only the one sideband used, while the numerator should include all sources giving rise to noise output, including noise converted from the unused sideband and arising from any spurious responses, resulting, for example, from local oscillator harmonics. Often, however, manufacturers quote noise figures for mixers which are 'broadband' or 'double sideband' figures, and incorporate, in the numerator, source thermal noise from unused mixer responses. In this case, the noise power in the denominator is higher, and therefore the quoted noise figure lower, than appropriate for normal operation. In fact, if there are no other spurious responses, the quoted double sideband noise figure will be half, or 3 dB lower, than the required single sideband figure.

7.3.6 Power meter noise correction

If the gain of the network under test is low, it may be that the noise of the power meter (and down-mixer, if used) is not negligible and a correction should be made for its influence on noise factor measurement. It is important to realise that it is not correct to note the power meter indication when it has a passive source impedance equal to that of the output impedance of the network under test and to subtract this from subsequent readings, since the thermal noise in this impedance is part of the noise power delivered by the network. Nor should a base power reading be derived by shorting the input of the power meter since this eliminates the effect of the equivalent input noise current generator. A possible method is to note the power meter indication when connected to a passive impedance equal to the output impedance of the network under test, subtract from this the calculated thermal noise from the impedance, and subtract this resulting power from subsequent readings. However, if it is possible to deal in terms of available powers, an easier method makes use of the expression for the noise figure of cascaded networks (equation (5.59)). Then the uncorrected measured noise factor is:

$$F_m = F + \frac{F_{pm} - 1}{G_a} \qquad (7.65)$$

where F and G_a are the noise factor and available gain of the network under test and F_{pm} is the noise factor of the power meter. Thus:

$$F = F_m - \frac{F_{pm} - 1}{G_a} \qquad\qquad (7.66)$$

and a correction to F_m can be made if F_{pm} and G_a are known. The available power gain of the network is easily measured as part of the noise factor measurement procedure and F_{pm} may be measured by treating the noise meter itself as the device under test. Other more sophisticated noise measurement systems will make this correction automatically from stored values of F_{pm} across the frequency range of the meter.

7.3.7 Comparison of sine-wave and noise generator methods

The main advantage of the noise generator method is that a knowledge of the network or measurement noise bandwidth is not required. At high frequencies, where a relatively wide measurement bandwidth (and therefore low averaging time-constant) may be employed, the output power doubling method is a fairly rapid method of measuring the noise figure. However, at low frequencies, or more specifically, narrow bandwidths, the long integration or averaging time required means that the high-signal sine-wave method is more rapid – particularly if the noise bandwidth is known – since only one noise measurement is required. The power doubling method is practically unusable for narrow bandwidths because of the time required to adjust the signal generator output to give a particular output power indication. In addition, it is more difficult to achieve a uniform spectral density from a noise generator at low frequencies as a result of $1/f$ noise contamination. There are non-uniformity problems at high radio frequencies arising, for example, from stray capacitance and inductance in the generator and connections. However, relatively small fractional bandwidths can be used at higher frequencies for spot frequency measurements. For example, a 10 kHz bandwidth, requiring an averaging time of only a few tens of milliseconds, represents a fractional bandwidth of only 1 per cent at 1 MHz. This means that the variation in spectral density across the measurement bandwidth is usually negligible, and the effect of generator spectrum non-uniformity can be more easily taken into account by using the spectral density appropriate for each spot frequency in the calculation of noise factor. Again, the more sophisticated noise factor measurement systems make this correction automatically, making use of stored values of the generator spectral density at spot frequencies.

The sine-wave method has the advantage that the equipment is usually more easily available and with narrow noise bandwidths there is a measurement time advantage in making only one noise power measurement. In addition, as we shall see later in the section on measurement tips, it is possible to reduce the problem of external interference arising as a result of the signal generator connection by replacing the signal generator by a shielded passive network with the same output impedance (this will often simply be a resistor), while making the noise measurement and measuring gain and bandwidth at a sufficiently high level to swamp the interference.

7.4 Noise temperature measurement

The noise temperature of a two-port network may be calculated from the noise factor using the relationship expressed in equation (5.36). Thus:

$$T_e = T(F - 1) \qquad (7.67)$$

or:

$$T_e = 290(F - 1)$$

if F is measured at 290 K.

If the noise temperature rather than the noise factor is required, equations (7.35)–(7.66) can be expressed directly in terms of noise temperature. For example, the equivalent temperature for the Y (hot/cold source power ratio) method, using equation (7.59) is (Adler *et al.*, 1963):

$$T_e = \frac{T_1 - YT}{Y - 1} \qquad (7.68)$$

The noise temperature of a source or single-port network may be calculated using the same comparison method. For example, if the noise temperature of an antenna is required, then switching the input of a receiver of known noise temperature between the antenna and screened network with the same impedance and known temperature T we have, from equation (7.68):

$$T_a = (Y - 1)T_r + YT \qquad (7.69)$$

where T_a is the antenna noise temperature and T_r is the receiver noise temperature.

7.5 Noise figure variation with source impedance, and noise parameters

In general, the variation of noise figure with source impedance requires a number of measurements with different values of Z_s (figure 7.14). At low frequencies with a resistive source, switching between a number of different source resistances will suffice. Note that a variable resistance should not be used because of the possibility of contact noise. As the frequency increases, stray capacitance and inductance (particularly in wire-wound resistors) may significantly change the source impedance. At high radio frequencies, the problem of stray reactive components becomes sufficiently severe that using the connecting leads to provide the correct impedance is advisable (Motorola Semiconductors, 1984). A quarter-wave length of transmission line (impedance Z_t) between signal generator (usually a noise generator at these frequencies) will transform the impedance (Z_g) of the signal generator to an impedance seen by the network under test of:

$$Z_s = \frac{Z_t^2}{Z_g} \tag{7.70}$$

Transmission lines of different impedances and lengths are required for this approach.

For resistive sources and uncorrelated equivalent input noise generators, the knowledge of only two noise parameters e_N and i_N, $\mathcal{E}_n$ and $\mathcal{I}_n$, R_{nv} and R_{ni} or F_{min} and R_{so} are required in order to calculate the noise figure at any source resistance (see section 5.2.9).

For resistive sources the minimum noise figure F_{min} and corresponding source resistance R_{so} are found by plotting F against R_s and determining the minimum.

Alternatively, the values of the equivalent input noise voltage and current sources can be measured. We have (from equation (5.21)):

$$e_{Ni}^2 = 4kTR_sB_m + e_N^2 + i_N^2R_s^2 \tag{7.71}$$

The value of e_N can be calculated by measuring e_{Ni} with a very low value of R_s such that e_N is very much greater than the other terms. Then:

$$e_N = e_{Ni} \tag{7.72}$$

The value of i_N can be measured by measuring e_{Ni} with a very high value of R_s such that $i_N^2R_s^2$ is very much greater than the other terms. Then:

$$i_N = e_{Ni}/R_s \tag{7.73}$$

A suitable check on whether the values of R_s are sufficiently small or high, respectively, is to double or halve the values and check that the calculated values of e_N and i_N change to a negligible degree.

At high frequencies, when the dependence of noise figure on source reactance is important (chapter 5, section 5.2.14), the source should have a variable reactive component. Variation of reactive impedance at a particular frequency may be achieved by a fixed inductor and variable capacitor, but the change in reactance across the measurement bandwidth should be acceptably small. At high radio frequencies, where lump resistance and reactive components are difficult to use, the reactive and resistive components of the impedance presented at the input of the network under test may be varied by the use of a transmission line with shorted stubs of adjustable position and length, or two fixed-position adjustable stubs (Haus *et al.*, 1960).

The minimum noise figure and optimum source impedance are found by adjusting both source resistance and reactance to give a minimum noise factor.

Where a non-zero source reactance is required for minimum noise figure, there are four network noise parameters to be determined, namely F_{min}, G_{so}, B_{so} and G_{nv} (equation (5.75)). These may be calculated by measuring the noise figures at four known source admittances and solving the four simultaneous equations, which can be written, using equation (5.75), for these unknown noise parameters. It may, in practice, be difficult to choose four values of Y_s to give a well conditioned set of equations, and an alternative is to make measurements at a large number of source admittance values and find the values of the noise parameters which give a least-squares best-fit to the measured noise figures (Lane, 1969; Kotyczka *et al.*, 1970).

7.6 Noise generators

Noise generators for noise figure measurements can take a number of different forms (figure 7.20). The simplest, as we have seen, is a resistor, the thermal noise from which is easily calculated and may be altered by changing its temperature. The shot noise current from a thermionic diode (Haus *et al.*, 1960), forward biased semiconductor diode or low voltage zener diode passed through a load resistance (Motchenbacher and Fitchen, 1973) gives a higher level output. The noise level can be increased further by using a reverse biased diode in the current limited avalanche breakdown mode (Kanda, 1976; Longley, 1978). At low frequencies, these active device noise generators may exhibit significant $1/f$ noise and it is important to choose devices which have $1/f$ noise corner frequencies (the frequency at which $1/f$ and thermal noise are equal) well below the lowest frequency of interest. At high radio frequencies, device capacitance and carrier transport time effects reduce the spectral density. Avalanche diodes specifically designed for noise generators are now standard sources at higher frequencies. At low frequencies, any of the above may be used provided that the $1/f$ corner frequency is sufficiently low. Even low noise amplifiers can be used as noise sources – low noise because they will have low popcorn and $1/f$ noise. Again, at low frequencies, the output of pseudo-random digital code

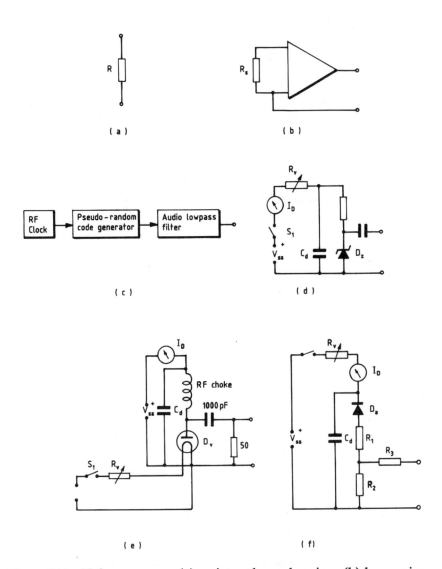

Figure 7.20 Noise sources: (a) resistor-thermal noise; (b) low noise
amplifier; (c) pseudo-random code generator; (d) zener diode
(D_z); (e) thermionic diode (D_v); (f) avalanche diode (D_a). Low
frequency (a)–(d), high frequency (e) and (f). In the diode
generators, R_v adjusts the diode current and S_1 switches the
noise source off. In (e), the output impedance changes little
between the 'on' and 'off' states, as it will in (f) with a suitable
choice of R_1, R_2 and R_3.

generators, low-pass filtered with a filter cut-off frequency well below the clock frequency, are suitable noise sources (Beastall, 1972; Horowitz and Hill, 1989). These sources will not necessarily have a Gaussian pdf and this may alter the measurement statistics.

Since it is important that the output impedance of the noise generator does not change significantly between the 'off' and 'on' states, it is necessary to buffer the impedance change in the active devices between these states using an output attenuator (figure 7.21).

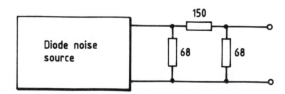

Figure 7.21 Attenuator used as impedance buffer. If the noise source 'on' to 'off' resistance changes from $15\,\Omega$ to $\infty\,\Omega$, the output impedance changes only from $47.9\,\Omega$ to $51.8\,\Omega$

We have seen how the output of the noise generator may be specified by means of its equivalent noise temperature T_{ne}. Another way of quoting the output is using the excess noise ratio (ENR) which is defined as (Longley, 1978):

$$ENR = 10\log_{10}\left(\frac{T_{ne} - 290}{290}\right) \quad \mathrm{dB} \tag{7.74}$$

This leads to a simplification of the calculation of noise figure (in dB) using the hot/cold source method. From equation (7.59), if $T = 290\,\mathrm{K}$, and identifying T_1 with T_{ne}, we have:

$$NF = 10\log_{10}\left(\frac{T_{ne} - 290}{290}\right) - 10\log_{10}(Y - 1) \quad \mathrm{dB}$$

$$= ENR - 10\log_{10}(Y - 1) \quad \mathrm{dB} \tag{7.75}$$

7.7 Active devices

Active devices require connection to supplies, which should be variable in order to be able to measure noise parameters under different operating conditions. It is obviously important that the supply circuits, particularly those setting input bias, either have an insignificant effect on the measured parameters or are taken into account when calculating the effective source and load conditions. For example, the simple arrangement shown in figure 7.22 may be used to measure the low frequency noise figure of a transistor and its variation with frequency and collector current.

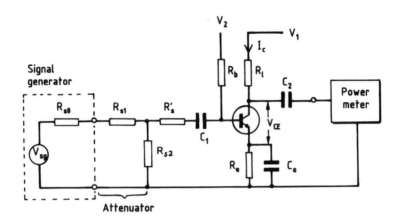

Figure 7.22 Circuit for the measurement of the low frequency noise of a bipolar transistor

At radio frequencies, the bias resistor may be replaced by an rf choke. Device noise figures are often measured with input and output reactances tuned out and with anti-phase capacitative feedback to neutralise the effect of collector–base capacitance (Motorola Semiconductors, 1984).

Refering to figure 7.22, supplies V_1 and V_2 are adjusted to set I_c and V_{ce} (the latter may be calculated from I_c, V_1 and R_l and R_e). The capacitors must all have negligible reactance for the lowest frequency under consideration. All the resistors should exhibit low excess noise. The value of R_{s2} should preferably be negligible compared with R'_s, and R_b should be very much greater than R'_s. In this case, the source resistance is:

$$R_s \simeq R'_s \tag{7.76}$$

In general, however, it is often difficult to maintain these conditions over a wide range of R_s values and we need to take into account the effect of R_b on R_s and the effective signal source voltage V_s. The required equivalence is shown in figure 7.23. Then:

$$R_s = (R_x + R'_s)//R_b \tag{7.77}$$
$$R_x = (R_{s0} + R_{s_1})//R_{s2} \tag{7.78}$$

and:

$$V_s = \frac{V_{sg} R_{s_2}}{R_{so} + R_{s_1} + R_{s_2}} \frac{R_b}{R_x + R'_s + R_b} \tag{7.79}$$

where R_{so} is the output resistance of the signal generator, V_{sg} is its unloaded signal voltage and $R_a//R_b$ indicates the value of R_a and R_b in parallel.

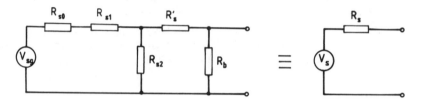

Figure 7.23   Device input circuit (figure 7.22) and its equivalent

A similar circuit may be used for measuring the low frequency noise parameters of FETs. The components R_e and C_e may be zero and the bias resistance R_b will be very high. The optimum source resistance may well be a few tens of megohms for MOSFETs and we would want to measure the noise factor at R_s values comparable and higher than this. Measurements of the small equivalent input noise current can present a problem – particularly in MOSFETs. The method described in section 7.5 requires the increase of the source resistance until the $i_N^2 R_s^2$ term is at least equal to the thermal noise and preferably significantly greater. However, this is difficult to achieve if i_N is very low. If, for example, i_N is 10^{-14} A in a bandwidth of 10 Hz, then a value of R_s as high as $10^8\,\Omega$ leads to a value of $i_N R_N$ of only 1 µV compared with a thermal noise voltage of 4 µV. If much higher resistances are used it becomes difficult to be sure of the true source impedance since small stray capacitance become significant even at low frequencies. For example, 1 pF has a reactance of $1.6 \times 10^9\,\Omega$ at 100 Hz. A solution to this problem which has been tried at low frequencies is to replace the source resistance with a small very low loss capacitor of reactance X_c (van der Ziel, 1976). Since the capacitor generates no thermal noise we have:

$$e_{Ni}^2 = e_N^2 + i_N^2 X_c^2 \tag{7.80}$$

or:

$$i_N = \frac{\sqrt{(e_{Ni}^2 - e_N^2)}}{X_c} \tag{7.81}$$

If we consider a capacitor of 100 pF and a measurement at 100 Hz at which frequency X_c is 16 MΩ then the noise current of 10^{-14} A considered earlier leads to a value of $i_N X_c$ of 1.6×10^{-7} V whereas a typical value of e_N in a bandwidth of 10 Hz is 10^{-8} V.

The method is really only usable at frequencies of the order of 100 Hz or less, since the very low capacitance values required at higher frequencies become comparable with stray capacitance. Another method, again only applicable at low frequencies where shot noise in the gate leakage current predominates, is to measure the leakage current using an electrometer and calculate the shot noise using equation (4.26). In the radio frequency region, where i_N is higher and increasing with frequency as a result of the channel thermal noise term (equation (6.55)), similar measurement techniques to those used for bipolar transistors can be used. There is a frequency range for which direct measurement of i_N is difficult and is best estimated by extrapolation of the low and high frequency measurements (figure 7.24) (Motchenbacher and Fitchen, 1973).

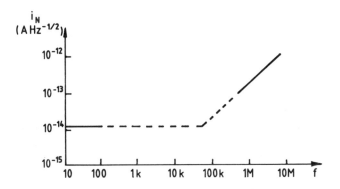

Figure 7.24 Estimation of FET equivalent input current noise by interpolation: ——— measured noise current; − − − estimated noise current

The noise parameters of operational amplifiers with high open loop gain are best measured with feedback applied to give a fixed gain and to stabilise the DC conditions. If the feedback resistors cannot be made small enough for their effects on the noise factor to be negligible, then the relationships developed in chapter 5, section 5.2.11 must be used to calculate the amplifier equivalent input noise current and voltage. It should be remembered that the common mode input voltage will alter the operating conditions of a differential amplifier and therefore its noise performance. If a differential amplifier is to be used with a significant common mode input, then the noise parameters should be measured under those conditions.

7.8 Very low frequency noise

At very low frequencies, noise is often referred to as 'drift' or 'base-line wander'. The designer is usually not so much interested in the rms level of the noise *per se*, but in the probability that the signal at these low frequencies remains within certain limits. For example, the noise of instrumentation amplifiers within a low frequency range (e.g. 0.0–1 Hz) may well be specified as having a 99.7 per cent probability of falling within certain limits. If the noise is Gaussian, then this is simply ±3 times the rms noise voltage in this frequency range. A way of checking this type of specification is to digitise the signal and calculate the proportion of the samples which fall outside the specified limits.

At even lower frequencies, the designer may well want to know how much the base line (short-term 'steady' output in the absence of a signal) changes over the duration of a measurement, or how often manual or automatic offset adjustments have to be made to bring the base line back to zero. The critical figure in this case is the change in base-line offset (referred to the input) over a specified time (for example, a minute or an hour). The specification in this case may be checked by sampling the amplifier output at the specified time interval and measuring the base-line change.

7.9 Excess noise in resistors

As mentioned in section 4.4, noise in excess of thermal noise is generated in resistors by the passage of a current. The normalised power spectrum has the form:

$$S_f(f) = K_f/f^\alpha \qquad\qquad (7.82)$$

where α, for resistors, has a value normally in the range 0.8–1.2 and is usually taken as unity. The basic measurement system required is shown in figure 7.25. With no current flowing in the resistor R, the measured normalised noise power is e_1^2 . This is the thermal noise in the resistor plus measurement system noise. With current I_{DC} flowing, the measured normalised power is e_2^2 and the measured excess noise is:

$$e_M^2 = e_2^2 - e_1^2 \qquad\qquad (7.83)$$

The excess noise is given by:

$$e_M^2 = \int_0^\infty \frac{K_f}{f} G_m(f)\mathrm{d}f \qquad\qquad (7.84)$$

where $G_m(f)$ is the normalised power gain of the filter and amplifier.

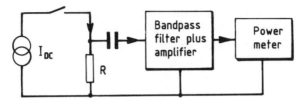

Figure 7.25   Measurement system for excess noise in resistors

If G_{po} is the mid-band gain of the filter, then this can be written:

$$e_M^2 = G_{po}K_fA \tag{7.85}$$

where:

$$A = \frac{1}{G_{po}} \int_0^\infty \frac{G_m(f)}{f} df \tag{7.86}$$

and is a measurable characteristic of the filter.

From equations (7.85) and (4.30), the excess noise voltage in one frequency decade is:

$$e_{FD} = e_M \sqrt{\left(\frac{\ln 10}{G_{po}A}\right)} \quad V_{rms} \text{ (frequency decade)}^{-1/2} \tag{7.87}$$

If the filter can be approximated by a rectangular bandpass filter with a passband from f_1 to f_2 then:

$$A = \frac{1}{G_{po}} \int_{f_1}^{f_2} \frac{G_{po}}{f} df$$
$$= \ln(f_2/f_1) \tag{7.88}$$

and:

$$e_{FD} = e_M \sqrt{\left(\frac{\ln 10}{G_{po}\ln(f_2/f_1)}\right)} \quad V_{rms} \text{ (frequency decade)}^{-1/2} \tag{7.89}$$

The excess noise parameter C_F (equation (4.34)) is:

$$C_F = e_{FD}/V_{DC}$$
$$= e_{FD}/(I_{DC}R) \quad V_{rms}V_{DC}^{-1} \text{ (frequency decade)}^{-1/2} \tag{7.90}$$

The choice of measurement bandwidth is determined by two constraints. A short measurement time for a given accuracy requires a wide bandwidth (section 7.2.4). However, thermal noise increases more rapidly with bandwidth than $1/f$ noise, leading to a reduced accuracy as the fractional difference between e_1^2 and e_2^2 (equation (7.83)) diminishes. A typical compromise is measurement in a 1 kHz band extending from 0.5 to 1.5 kHz. This measurement passband and a CR averaging time constant of 1 second leads to an rms random power measurement error of about 2 per cent.

We define the sensitivity of the instrument as the value of C_F that gives an excess noise equal to the thermal noise for a given power dissipation in the resistor. Under these conditions:

$$4kTR(f_2 - f_1) = K_f \ln(f_2/f_1) \tag{7.91}$$

Eliminating e_{FD} from equations (4.30) and (4.34) and substituting for K_f in the above leads to:

$$C_F = \left[\frac{4kT(f_2 - f_1) \ln 10}{W \ln(f_2/f_1)} \right]^{1/2} \tag{7.92}$$

where W is the power dissipated in R. If W is limited to 0.1 watt to avoid significant heating of R, T is 290 and, as above, $f_1 = 0.5$ kHz and $f_2 = 1.5$ kHz then:

$$C_F = 0.0183 \ \mu V \ V_{DC}^{-1} \ (\text{frequency decade})^{-1/2} \tag{7.93}$$

which corresponds to a noise index of -35 dB. Note that we have assumed negligible noise contribution from the measurement system. In practice, it would be difficult to maintain this condition over a wide range of resistor values. In addition, as the excess noise becomes comparable with thermal and measurement system noise, the measurement random error becomes significantly worse than that calculated above since we have to calculate the excess noise from the difference of two comparable random quantities (equation (7.83)).

It is obviously important that the current source (figure 7.25) is low noise and, specifically, that the noise current from the source gives rise to a noise voltage drop across R which is small compared with the thermal noise. If the noise current is known then it is possible to subtract the noise power from this source from the normalised measured excess noise but, of course, the sensitivity is reduced. In addition, real current sources have a finite parallel impedance and this should be very much greater than the maximum resistance under test.

An alternative (Conrad *et al.*, 1960) to the current source is the arrangement shown in figure 7.26, which uses a voltage supply and an isolation resistance R_m which should be comparable or greater than R in order to avoid high attenuation of the noise from R. A calibration signal generated across a resistor which is negligibly small compared with R removes the necessity of calculating the attenuation resulting from the potential divider formed by R and R_m. With V_{DC} reduced to zero (note that this must not be achieved by an open circuit – R_m plus the internal impedance of the supply must remain approximately constant), the system gain G_{po} is measured using the calibration signal (very much greater than the thermal noise). This is followed by a measurement of thermal noise (calibration signal switched off) followed in turn by a measurement of the thermal plus excess noise with V_{DC_1} set to give the desired voltage drop V_{DC}. The calculations are the same as those using the current source. Note that both the noise from the supply and the excess noise in the isolation resistance should be negligible compared with the thermal noise in $R//R_m$. A check on this condition may be carried out by putting a resistor of similar construction and value to the isolation resistor (very low excess-noise, wire-wound) in the test position (R) and noting any increase in measured noise when the DC supply is switched on.

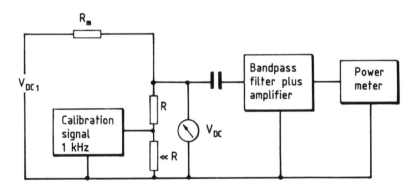

Figure 7.26 Measurement system for excess noise in resistors using direct voltage source and calibration signal

In the above analyses, we have assumed α to be unity. It is worth briefly investigating the effect of variations from this value, on a calculation of excess noise in a particular frequency band from the measurement of noise voltage per root frequency decade.

To simplify the analysis we assume rectangular passbands and refer measurements to the input of the measurement system.

If α is not equal to 1 and the excess noise measurement is carried out in a frequency range from f_{m_1} to f_{m_2} then the measured normalised excess noise power is:

$$e_{Fm}^2 = \int_{f_{m_1}}^{f_{m_2}} \frac{K_f}{f^\alpha} df$$

$$= \frac{K_f}{\alpha - 1} \left(\frac{1}{f_{m_1}^{\alpha-1}} - \frac{1}{f_{m_2}^{\alpha-1}} \right) \tag{7.94}$$

and the normalised excess noise power per frequency decade calculated from this on the assumption that $\alpha = 1$ is (using equation (7.89)):

$$e_{FDM}^2 = \frac{K_f}{\alpha - 1} \left(\frac{1}{f_{m_1}^{\alpha-1}} - \frac{1}{f_{m_2}^{\alpha-1}} \right) \frac{\ln 10}{\ln(f_{m_2}/f_{m_1})} \tag{7.95}$$

If we then use this figure to calculate the excess noise in a frequency band f_1–f_2 (by multiplying by the number of frequency decades in this range), this excess noise is:

$$e_c^2 = \frac{K_f}{\alpha - 1} \left(\frac{1}{f_{m_1}^{\alpha-1}} - \frac{1}{f_{m_2}^{\alpha-1}} \right) \frac{\ln 10}{\ln(f_{m_2}/f_{m_1})} \log_{10}(f_2/f_1)$$

$$= \frac{K_f}{\alpha - 1} \left(\frac{1}{f_{m_1}^{\alpha-1}} - \frac{1}{f_{m_2}^{\alpha-1}} \right) \log_{10}(f_2/f_1) / \log_{10}(f_{m_2}/f_{m_1}) \tag{7.96}$$

The actual noise in this frequency band is:

$$e_a^2 = \frac{K_f}{\alpha - 1} \left(\frac{1}{f_1^{\alpha-1}} - \frac{1}{f_2^{\alpha-1}} \right) \tag{7.97}$$

and the fractional error in the calculated noise rms voltage is:

$$\varepsilon_F = \frac{e_c}{e_a} - 1$$

$$= \left[\left(\frac{f_2}{f_{m_2}} \right)^{\alpha-1} \frac{\left((f_{m_2}/f_{m_1})^{\alpha-1} - 1 \right)}{\left((f_2/f_1)^{\alpha-1} - 1 \right)} \frac{\log(f_2/f_1)}{\log(f_{m_2}/f_{m_1})} \right]^{1/2} - 1 \tag{7.98}$$

As might be expected, the error increases with the deviation of α from unity, is zero if the measurement and design frequency bands coincide, and increases with increasing difference between the geometric bandwidths f_{m_2}/f_{m_1} and f_2/f_1 and with increasing difference between their upper frequency limits.

As an example, we consider measurements in the frequency band 0.5–1.5 kHz and the calculation from this, of the excess noise in an audio bandwidth from 20 Hz to 10 kHz. In this case, $\varepsilon_F = -9.2$ per cent if $\alpha = 1.2$, or 3.6 per cent if $\alpha = 0.8$.

If, however, the design frequency band is significantly different from the measurement band, for example, f_1 is 1 Hz and f_2 is 100 Hz, then the corresponding errors are -37 per cent and 54 per cent.

7.10 Measurement tips

Some of the problems of noise parameter measurement and methods of overcoming them have been covered in previous sections. This section will deal with some more general points.

One major problem in noise parameter measurement is the avoidance of interference. Thermal noise signals are small and it is very easy for comparably small signals from other circuits to be coupled into the noise signal path and give erroneous readings. The signal generator and its connections may introduce interference by acting as an antenna, by capacitative and inductive coupling particularly from the power-line cables and, if connected to the power-line ground, by common signal path (ground loop). If the signal generator is used at a fairly high-level output, the interference may be negligible. However, it is not always possible to operate at a signal level which is sufficiently large to avoid interference problems but sufficiently small to avoid device non-linearity. The power doubling method of noise factor measurement requires small signal levels. Even if the signal generator is switched off when making the thermal noise measurement, interference may still enter by the same pathways.

Battery-powered signal generators avoid interference arising from the power-line or ground loop but still allow electromagnetic, inductive and capacitative interference coupling. The problem can be reduced if the signal generator can be used at a high level for gain measurement, or as a 'hot' source, and then be disconnected and replaced by a screened passive one-port network having the same output impedance as the signal generator, for the thermal noise measurement.

High-source resistances are often used at low frequencies and present particular problems. Firstly, a high-source impedance together with a high amplifier input impedance is especially susceptible to electromagnetic and capacitatively coupled interference, and good screening is essential.

Secondly, the resistance of high resistors can be significantly lowered by moisture or other surface contamination providing a second electrical pathway between the end connections. Thirdly, stray capacity shunting the source resistor can significantly reduce the impedance of the resistor, leading to an increase in gain and a reduction in thermal noise.

At high radio frequencies, the normal care to avoid impedance mismatches in a nominally matched connection should be taken. Dirty or worn connectors can be a cause of this, and it is worth pointing out that the plating on well used connectors does eventually wear out, requiring the connector to be replaced.

References

Adler, R. *et al.* (1963). 'Description of the noise performance of amplifiers and receiving systems', *Proceedings of the IEEE*, **51**, 436–442.

Baxandall, P. J. (1968). 'Noise in transistor circuits, Part 2', *Wireless World*, **74**, 454–459.

Beastall, H. R. (1972). 'White-noise generator', *Wireless World*, **78**, 127–128.

Bendat, J. P. and Piersol, A. G. (1971). *Random Data: Analysis and Measurement Procedures*, Wiley, New York.

Blachman, N. M. (1966). *Noise and its Effect on Communication*, McGraw-Hill, New York.

Bracewell, R. N. (1986). *The Fourier Transform and its Application*, 2nd edn, McGraw-Hill, New York.

Conrad, G. T., Newman, N. and Stansbury, A. P. (1960). 'A recommended standard resistor-noise test system', *IRE Transactions on Component Parts*, **CP-7**, 71–88.

Davenport, W. B. and Root, W. L. (1958). *Random Signals and Noise*, McGraw-Hill, New York.

Harman, W.W. (1963). *Principles of the Statistical Theory of Communications*, McGraw-Hill, New York.

Haus, H. A. *et al.* (1960). 'IRE standards on methods of measuring noise in linear twoports, 1959', *Proceedings of the IRE*, **48**, 60–68.

Hewlett Packard (1983). 'Fundamentals of RF and microwave noise figure measurements', *Application Note 57-1*.

Hewlett Packard (1988). 'Noise figure measurement accuracy', *Application Note 57-2*.

Horowitz, P. and Hill, W. (1989). *Art of Electronics*, 2nd edn, Cambridge University Press.

Kanda, M. (1976). 'An improved solid-state noise source', *IEEE Transactions on Microwave Theory and Techniques*, **MTT-24(12)**, 990–995.

Kotyczka, W., Leupp, A. and Strutt, M. J. O. (1970). 'Computer aided determination of two-port noise parameters (CADON)', *Proceedings of the IEEE*, **58**, 1850–1851.

Lane, Q. R. (1969). 'The determination of device noise parameters', *Proceedings of the IEEE*, **57**, 1461–1462.

Lathi, B. P. (1968). *An Introduction to Random Signals and Communication Theory*, International Textbook Co, Scranton, Pennsylvania.

Longley, S. R. (1978). 'Design and application information for broadband solid-state noise sources', *Mullard Technical Communications*, **138**, 309–316.

Middleton, D. (1960). *An Introduction to Statistical Communication Theory*, McGraw-Hill, New York.
Motchenbacher, C.D. and Fitchen, F.C. (1973). *Low-Noise Electronic Design*, Wiley, New York.
Motorola Semiconductors (1984). *RF Device Data Book*, pp. 8.17–8.27.
Papoulis, A. (1984). *Signal Analysis*, McGraw-Hill, New York.
van der Ziel, A. (1976). *Noise in Measurements*, Wiley, New York.

8   Computer Modelling

8.1   Introduction

Computer Aided Design, involving simulation of circuit operation, is a powerful tool for the circuit designer. It is almost impossible to design any but the simplest circuit without overlooking some design point which is important to correct operation. Testing, by a computer simulation, a design, rather than building and testing, enables far more rapid progress towards the final circuit. Of course, all simulators involve approximations in device modelling and, except in simulators specifically designed for microwave frequencies or when specifically included by the designer (for example, Hartmann and Strutt, 1974), take no account of parasitic components (e.g. PCB track resistance and inductance, or capacitance and mutual inductance between tracks) resulting from interconnections, and the final prototypes must be real rather than simulated.

In low signal-level circuits it is important that the simulation should include noise sources so that noise performance and particularly the effect of design changes, dictated by other performance constraints, may be checked. As shown earlier, a critical section of the circuit in low noise design is the source and first amplifier. If the designer is frequently concerned with system input circuits having the same basic structure, it may be worthwhile simply writing a program which calculates the noise parameters of interest from inputs of those circuit parameters which change between designs. Alternatively, the designer can make use of published programs for particular types of circuit – for example, for sensors and their preamplifiers (Motchenbacher and Fitchen, 1973).

However, there are a number of advantages in using the noise analysis facility in a general circuit simulator such as SPICE. This has the flexibility to permit the analysis of any circuit configuration and a noise analysis along with analyses of other circuit functions. In many circuit design requirements, noise level or signal-to-noise ratio is not an overriding concern, but has a certain weight along with many other requirements (frequency response, gain, tolerance of circuit element variation etc.). In this case, it is clearly advantageous for the designer to work with simulation software which provides information on all the circuit characteristics. In addition, industry-standard software such as SPICE with its standard format for circuit description files allows easy transfer of information between people and sites.

8.2 SPICE

8.2.1 Introduction

The circuit simulation program SPICE (Simulation Programme with Integrated Circuit Emphasis) was developed at the University of California, Berkeley during the 1970s for use on mainframe computers, and, in its original form, is in the public domain. There are now many commercial versions of SPICE which are sold together with support software giving improved output file handling and display, and sometimes improved simulation, for use on both mainframe and personal computers. Their utility has been increased by the provision of SPICE format circuit description files as optional outputs from schematic design software packages.

It is not appropriate to describe the operation or use of SPICE in detail here, as there are many texts on the subject (for example, Meares and Hymowitz, 1988; Tuinenga, 1988; Rashid, 1990) and commercial versions of SPICE come with their own user manuals. The description here will be brief and concentrate on its use in noise analysis.

8.2.2 Input file

The program requires, as an input, a description of the circuit to be simulated plus the type of analysis and output required. It has the form shown in table 8.1.

Table 8.1 SPICE input file format

Title line
Circuit description
Analysis description
Output description
Models
End statement

The order of the file between the title line and end statement may be altered.

The circuit is described in SPICE by reference to numbered nodes (junctions between circuit elements) and the circuit description is simply a list of circuit elements together with the nodes to which they are connected. Each line has the format:

name   1st node   2nd node   . . . .   value

the order of the nodes obviously being important with elements having polarity (e.g. diodes) and elements having more than two connections (e.g. bipolar transistors, FETs). Circuit elements are distinguished by the initial letter of their name, as listed in table 8.2.

Table 8.2   Initial letter of circuit element name

C	Capacitor
D	Diode
E	Voltage-controlled voltage source
F	Current-controlled current source
G	Voltage-controlled current source
H	Current-controlled voltage source
I	Independent current source
J	Junction FET
L	Inductor
M	MOSFET
Q	Bipolar transistor
R	Resistor
V	Independent voltage

For example, the SPICE circuit description of the circuit shown in figure 8.1 is as follows:

```
VCC  4  0   15V
VIN  0  1   AC   100MV
RS   1  2   1K
C1   2  3   1UF
R1   3  0   30K
R2   3  4   120K
Q1   5  3   6   QN4124
R3   0  6   2.2K
C2   0  6   100UF
R4   4  5   5.6K
C3   5  7   1UF
R5   0  7   10K
```

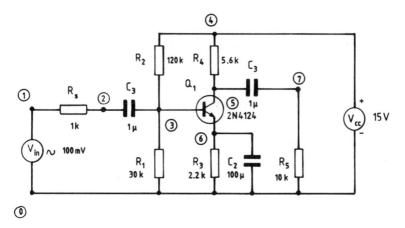

Figure 8.1   Single-stage amplifier

8.2.3   *Analyses and output*

SPICE can be instructed to perform a number of analyses of the described circuit. These include: the calculation of the bias points or steady voltage levels at each node; the variation of node voltages when the value of an input voltage or current source, a model parameter or the temperature is swept through a range of values; the frequency response; the time-variation of the voltages at the nodes in response to a particular input waveform (transient analysis); and the calculation of the noise voltage at a specified node, and that noise referred to a specified voltage or current generator (usually the input signal generator). The current in any connection in the circuit can be displayed in all these analyses by including a zero-voltage generator in that connection and asking for the current flowing through the generator. Some versions of SPICE allow the display of current through specified resistors. Of particular interest here, since they are both required for noise analysis, are the calculations of bias points (DC analysis) and the frequency response (AC analysis). The bias points are in fact calculated before any of the other analyses, since the operating conditions of many of the circuit elements are determined by the steady-state conditions. The DC analysis is not considered separately here.

AC analysis

The statement:

.AC   x   NP   FSTART   FSTOP

initiates the frequency response analysis. FSTART and FSTOP specify the frequency range; x is either LIN, OCT or DEC specifying respectively a linear, $\log_2$ or $\log_{10}$ frequency sweep; and NP is the number of frequencies, within the range of a linear sweep or per decade (octave), at which the response is calculated.

The output commands .PRINT and .PLOT enable the output of the analysis to be printed or plotted on a line printer. The commands:

.PRINT   AC   VM(7)   VP(7)

and

.PLOT   AC   VM(7)   VP(7)

for example, give a printout and a plot of the magnitude (VM) and phase (VP) of the signal at node 7 (with respect to ground) as the frequency of the source is varied as specified in the .AC statement.

Commercial versions of SPICE have graphics software which allows the screen display of node voltages and currents, and the option to obtain hardcopy on a laser printer or an x–y plotter.

Noise analysis

The noise analysis calculates the contribution from each independent noise source in the circuit to the noise voltage at a specified output node, making use of a computed frequency transfer function between each source and the output. The total output noise voltage is calculated as the square root of the sums of the squares of these contributions. The equivalent input noise voltage (or noise current if the input source is a current generator) is then calculated using the computed frequency-dependent gain between the input and output nodes. The noise analysis statement has the form:

.NOISE   V(output node no.)   input source name   interval

The interval (an integer n), if present, tells SPICE to print out the individual contributions to the total output noise every nth frequency specified in the .AC statement.

As in the frequency response analysis, the noise analysis results may be printed or plotted on a line printer, but it is usually preferable to make use of the graphics software supplied.

The noise output statements have the forms:

.PRINT   NOISE   INOISE   ONOISE
.PLOT   NOISE   INOISE   ONOISE

The noise analysis results are the root normalised power spectrum in $V Hz^{-1/2}$, at the output node (ONOISE) or input source (INOISE), or on a dB scale referred to $1 V Hz^{-1/2}$. The output software will often allow the designer to integrate the noise spectrum (remembering to square to obtain the normalised power spectrum before integrating) in order to obtain the total noise power.

Note that some versions of SPICE calculate the equivalent input noise by dividing by the output voltage derived from the .AC analysis rather than the gain – assuming an input signal of 1 V (or 1 A if a current generator is used). In this case, INOISE should be multiplied by the input generator voltage (or current) in order to obtain the true equivalent input noise.

The plots of voltage gain (V(7)/Vin) and input and output noise (root normalised spectral density) against frequency for the circuit shown in figure 8.1 are given in figure 8.2(a). The corresponding plots for source resistances of $10 k\Omega$ and $100 k\Omega$ are shown in figures 8.2(b) and 8.2(c). The total output noise V_o is obtained by integrating the square of the output noise expressed in $V Hz^{-1/2}$ and taking the square root of the result. The equivalent input noise V_i is obtained by dividing this figure by the mid-band voltage gain. Note that integration, by summing the noise spectral density at the frequency intervals specified in the .AC statement and multiplying by the interval, requires a linear frequency sweep.

8.2.4 Device models

Introduction

The device models used in the SPICE simulation are simply mathematical descriptions of the electrical characteristics of the devices. There may be a number of options for the models of semiconductor devices which follow different published models. The noise models are essentially those described in chapter 6. For example, the noise model of a resistor is shown in figure 8.3. Note that a parallel noise current rather than a series voltage generator is used in order to avoid extra nodes. It is, however, important to determine the limitations of the noise modelling in the SPICE version used. For example, the input current noise in FETs is often not modelled, and therefore the capacitatively coupled channel thermal noise which may have a significant effect at high frequencies will not appear.

Model parameters which distinguish one device type from another (a 2N2906 from a 2N4403 transistor, for example) may be set by the user.

.MODEL statement and libraries

Each active device in the circuit description must refer to a .MODEL statement which tells SPICE the properties of the device. The statement has the form:

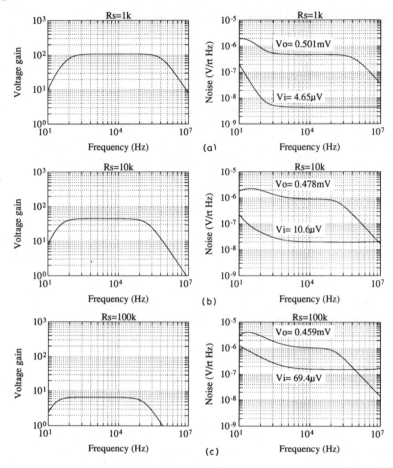

Figure 8.2   Frequency response and noise for circuit in figure 8.1. Output and equivalent input noise are V_o and V_i respectively. (a) Source resistance equals $1\,k\Omega$; (b) source resistance equals $10\,k\Omega$; (c) source resistance equals $100\,k\Omega$

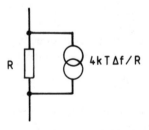

Figure 8.3   SPICE noise model of a resistor

.MODEL   model name   type reference (parameters)

The model type reference is shown in table 8.3.

Table 8.3   Model type reference

D	Diode
NPN	*NPN* bipolar transistor
PNP	*PNP* bipolar transistor
NJF	*N*-channel JFET
PJF	*P*-channel JFET
NMOS	*N*-channel MOSFET
PMOS	*P*-channel MOSFET

If no parameters are specified, then default values are taken. For example if an *NPN* bipolar transistor Q1 is required with a base resistance of $100\,\Omega$ (default zero) and flicker noise coefficient of 3.2×10^{-16} (default zero), then the circuit description and model statement would have the form:

Q1   1   2   3   QMOD1
.MODEL   QMOD1   NPN   (RB = 100 KF = 3.2E-16)

Alternatively, the device libraries, in which the parameters for a large number of devices are stored, may be used. In this case, the device is simply specified by name (see the description of Q1 in figure 8.1), but a library reference statement (format dependent on SPICE version) must be included.

Operational amplifiers

Operational amplifiers could be modelled by means of their circuits. However, since the SPICE analysis time depends on the number of nodes this results in very long analysis times. Instead, macro-models making use of a small number of components together with controlled voltage and current sources are often used. For example a very simple model of an op-amp with a single pole bandwidth defining the filter $(C_1 R_1)$, voltage gain 1000 and input and output resistances R_{in} and R_{out} is shown in figure 8.4.

However, since there are no active devices in this circuit and the noise is defined solely by thermal noise in the resistors, the noise performance will be quite misleading. A more complex macro-model with a bipolar transistor input stage is shown in figure 8.5.

Library files of macro-models of commercial operational amplifiers are available with SPICE software, and the models may be called by their device name.

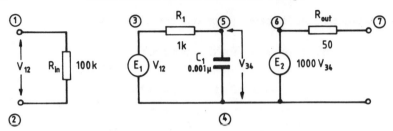

Figure 8.4   Simple SPICE op-amp model

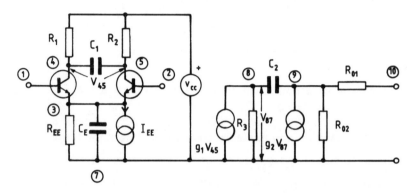

Figure 8.5   Bipolar transistor input SPICE op-amp model

8.2.5   Sub-circuits

Macro-models and other circuit blocks that the user wishes to use in a number of circuits or positions in one circuit are stored and called sub-circuits. These sub-circuits may be written into an input file or called from a library.

For example, the sub-circuit description of figure 8.4 is:

```
*SUBCIRCUIT, TYPE, INPUT NODES, GROUND, OUTPUT NODE
.SUBCKT      OPAMP1    1 2            4               7
RIN  1  2  100K
E1  3  4  1  2  1
R1  3  5  1K
C1  5  4  0.001UF
E2  6  4  5  4  1000
ROUT  6  7  50
.ENDS
```

Note the use of an asterisk to indicate a comment line, the form of the .SUBCKT line incorporating the name of the subcircuit (OPAMP1) and the connection nodes, and the format of the controlled source (E1 and E2) statements:

source name   source nodes   control nodes   multiplier

As an example, the statement to insert this operational amplifier sub-circuit in a main circuit file, where the amplifier is called XAMP, the inputs are connected to nodes 20 and 21, the ground to 0 and the output to 22, is:

XAMP   20   21   0   22   OPAMP1

The 'X' identifies the device as an operator-defined sub-circuit.

8.2.6 Noise generators

It is possible to insert noise sources within a circuit by using controlled voltage and current sources to transfer noise from other devices – for example, thermal noise from resistors, or shot and flicker noise from diodes (Whalen and Paludi, 1977; Scott and Chen, 1987).

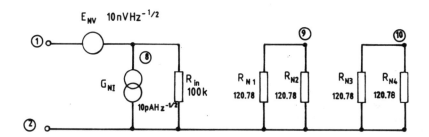

Figure 8.6   Resistor thermal noise used to create input noise generators

The simple operational amplifier macro-model may be improved by inserting parallel current and series voltage equivalent input noise generators as shown in figure 8.6. Note that the noise from these sources is in addition to the thermal noise in the resistors in figure 8.4.

The extra statements required are:

RN1   2   9   120.78
RN2   2   9   120.78
RN3   2   10   120.78
RN4   2   10   120.78
GNI   8   2   9   2   0.01
ENV   1   8   10   2   10

Two resistors in parallel are used for each thermal noise source, since SPICE requires two connections to each numbered node. The multipliers at the end of the GNI and ENV statements are used to convert the $1\,\mathrm{nV\,Hz^{-1/2}}$ noise of a $60.39\,\Omega$ resistance (two $120.78\,\Omega$ resistors in parallel) at $300\,\mathrm{K}$ to $10\,\mathrm{pA\,Hz^{-1/2}}$ and $10\,\mathrm{nV\,Hz^{-1/2}}$ respectively. Note the SPICE default temperature of $300\,\mathrm{K}$ (27°C) rather than $290\,\mathrm{K}$.

Note also that these are constant spectral density generators and that separate thermal noise sources should be used for each independent noise source inserted. The use of a single thermal noise source would lead to completely correlated noise generators.

Noise generators with frequency-dependent spectra may be modelled using more complex noise source circuits. For example, noise with a spectrum increasing at $6\,\mathrm{dB}$ per octave, which is often a good approximation for the high frequency spectra of equivalent input noise generators, may be generated using the circuit shown in figure 8.7.

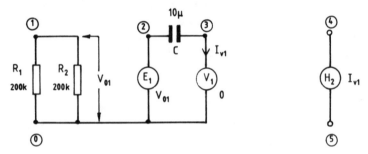

Figure 8.7   High frequency noise generator

The current I_{v_1} flowing through zero voltage generator V_1 is:

$$I_{v_1} = j2\pi f C V_{01} \tag{8.1}$$

where V_{01} is the thermal noise emf of $R\,(=R_1//R_2)$. The current controlled voltage generator H2 then has a noise root spectral density of $2\pi f C (4kTR)^{1/2}$ and may be connected in series with ENV in figure 8.6 to give a rising high frequency noise spectrum. The components C and R set the noise level.

The SPICE statements for the circuit in figure 8.7 are:

```
R1  0  1  200K
R2  0  1  200K
E1  0  2  0  1  1
C   2  3  1E-5
V1  0  3  DC  0
H2  5  4  V1  1
```

These values of C_1 and $R_1//R_2$ give a root spectral density of $2.56\,\mu V\,Hz^{-1/2}$ at 1 MHz. Note the format for the zero voltage generator V1 which is required for reference in the H2 statement. The numerical value of the voltage of H2 in volts is equal to that of the current flowing through V1 in amps. The multiplier has been set to unity. If a current generator is required, an F device is used in place of H2.

The flicker and shot noise current of a semiconductor junction may be used to create a noise generator with both constant spectral density and flicker noise components. For example, the circuit of figure 8.8 may be used in conjunction with an E or G source with output proportional to the voltage between nodes 1 and 2. Since the current from the independent direct current generators I1 and I2 are equal and the diodes DF1 and DF2 are similar, there is no steady voltage across nodes 1 and 2, only shot and flicker noise from the diodes. If the series parasitic resistance of the diode model is put equal to zero (usual default value), then the small signal AC resistance is:

$$r_e = kT/(eI) \tag{8.2}$$

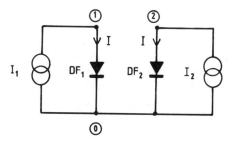

Figure 8.8 Noise generator with flicker and constant spectral density components

The shot and flicker noise current spectral densities for one diode are:

$$\mathcal{I}^2_{S_1} = 2eI$$

$$\mathcal{I}^2_{F_1} = K_F I/f \tag{8.3}$$

respectively.

Note that the exponent of I in the flicker noise expression is unity. This is the normal default value.

The noise voltage across each diode is calculated from the product of r_e and the noise current. The spectral density of the noise voltage across nodes 1 and 2 is double that arising from one diode and is:

$$V_{\text{Sn}}^2 = 4eI(kT/(eI))^2$$

or:

$$V_{\text{Sn}} = 2.07 \times 10^{-11} I^{-1/2} \quad \text{V Hz}^{-1/2} \tag{8.4}$$

for constant spectral density noise.

If I is 0.43 mA then:

$$V_{\text{Sn}} = 10^{-9} \quad \text{V Hz}^{-1/2} \tag{8.5}$$

The flicker noise corner frequency f_L is found by equating $\mathcal{I}_{\text{S}_1}$ and $\mathcal{I}_{\text{F}_1}$. Thus:

$$K_F = 3.2 \times 10^{-19} f_L \tag{8.6}$$

If f_L is 1 kHz:

$$K_F = 3.2 \times 10^{-16} \tag{8.7}$$

The required level of noise can be set using the multiplier of the E or G generator. If flicker noise only is required, then the multiplier should be set sufficiently low that the constant spectral density noise is negligible compared with existing noise, and K_F set to give the required flicker noise spectral density at 1 Hz.

A generator giving independently controlled flicker, constant density and 6 dB/octave high frequency (hf) noise is shown in figure 8.9. The source noise for the hf generator, in this case, is diode shot noise rather than resistor thermal noise.

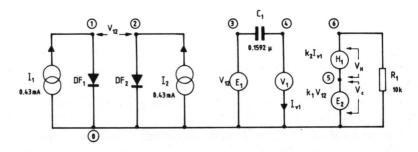

Figure 8.9   Flicker, constant density and high frequency noise generator

The flicker noise corner frequency is given by equation (8.6). The constant spectral density noise voltage is:

$$V_C = 10^{-9} k_1 \quad \mathrm{V\,Hz^{-1/2}} \tag{8.8}$$

and the high-frequency noise is:

$$V_H = k_2 2\pi C_1 \times 10^{-9} f$$

$$= k_2 \times 10^{-15} f \quad \mathrm{V\,Hz^{-1/2}} \tag{8.9}$$

For a high-frequency corner frequency of f_h MHz:

$$k_2 = k_1 / f_h \tag{8.10}$$

The output V(6) of a circuit with low and high corner frequencies of 1 kHz and 1 MHz respectively, and a constant spectral density noise voltage of $10^{-9} \mathrm{\,V\,Hz^{-1/2}}$, is shown in figure 8.10. The SPICE input file is shown below.

```
FLICKER, CONSTANT-SPECTRUM AND HF NOISE GENERATOR
I1  0  1  DC  0.43MA
I2  0  2  DC  0.43MA  AC  1MA
DF1  1  0  DFF1
DF2  2  0  DFF1
E1  0  3  1  2  1
C1  3  4  0.1592UF
V1  0  4  DC  0
E2  0  5  1  2  1
H1  5  6  V1  1
R1  0  6  10K
.MODEL  DFF1  D(KF=3.2E-16)
.AC  DEC  5  0.01HZ  100MEGHZ
.NOISE  V(6)  I2
.PRINT  AC  V(6)
.PRINT  NOISE  INOISE  ONOISE
.END
```

Note that R_1 has been included in order to test this circuit in isolation, and I_2 has been given an arbitrary AC component in order to have an 'input' source for INOISE. When used as a sub-circuit, R_1 and the AC component of I_2 will not be required, and E2 will not necessarily be grounded. For a corresponding noise current generator, H1 and E2 should be replaced by controlled current sources in parallel.

The use of a macro-model such as that shown in figure 8.5 may not be very accurate for noise modelling since the active device parameters are

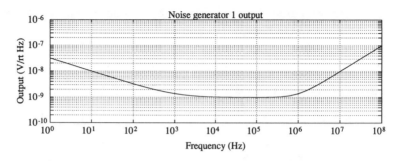

Figure 8.10   Noise output from circuit in figure 8.9

usually set values enabling accurate modelling of other performance characteristics (pulse response rise time, frequency response etc.), and may not be those giving an accurate model of noise performance. For example, many operational amplifiers have active device rather than resistive loads (the 741 op-amp, for example). In these devices, the equivalent input voltage generator may have a higher spectral density and/or flicker noise corner frequency as a result of the noise in the collector currents of the input transistors generated by the active loads. This extra noise (i_{Ne}) is equivalent to an input series noise voltage (i_{Ne}/g_m) – see chapter 6, section 6.8. The circuit in figure 8.8 may be used to drive a flicker, and constant density noise voltage generator added in series with the input of such a model in order to improve its noise modelling.

One of the limitations of SPICE is that the excess noise in resistors is not modelled. It is possible to use an arrangement similar to that described above to put a flicker noise current generator in parallel with a resistor, making sure that the shot noise component is small compared with the resistor thermal noise. A simpler method is to connect a reversed bias diode across the resistor and again adjust the value of K_F in the diode model to give the desired spectral density (Skrzypkowiak and Chen, 1987).

A serious limitation of the use of noise generators, used as described above, to improve SPICE noise modelling accuracy is that the generators have fixed outputs. If the DC conditions are changed, the generator outputs may have to be re-calculated.

It is, of course, possible to change the SPICE source code to improve noise modelling (Andrian *et al.*, 1987).

8.2.7   Noise analysis example

As an example of the use of SPICE to measure the noise performance, along with other characteristics, the fourth-order 1 dB-ripple Chebyschev filter shown in figure 8.11 has been modelled.

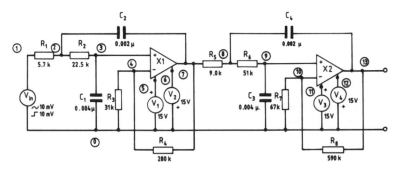

Figure 8.11 Fourth-order, 1 dB-ripple Chebyshev filter

The root spectral density of the input voltage and current generators of the µA741 operational amplifier model used are shown in figure 8.12. The SPICE input file with the circuit description and the statements required for transient (step) response, frequency response, output noise and noise referred to the input generator is shown below:

```
CHEBYSCHEV FILTER,  4POLE,  1DB RIPPLE
R1   1   2   5.7K
R2   2   3   22.5K
C1   3   0   0.004UF
C2   2   7   0.002UF
R3   4   0   31K
R4   4   7   280K
X1   4   3   7   5   6   UA741
R5   7   8   9.0K
R6   8   9   51K
C3   9   0   0.004UF
C4   8   13   0.002UF
R7   10   0   67K
R8   10   13   590K
X2   10   9   13   11   12   UA741
V1   5   0   15V
V2   0   6   15V
V3   11   0   15V
V4   0   12   15V
VIN   0   1   AC   10MV   PULSE   0   10MV
.TRAN   0.005MS   1MS
.AC   LIN   200   0.1KHZ   100KHZ
.NOISE   V(13)   VIN
.PRINT   AC   VM(13)
```

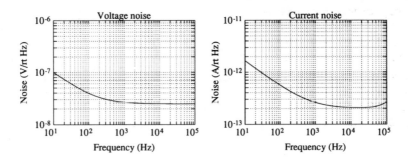

Figure 8.12 Input voltage and current noise of op-amps used in the circuit of figure 8.11

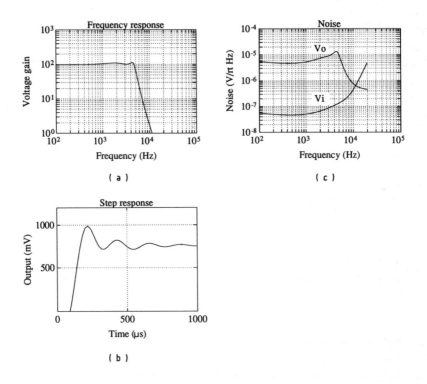

Figure 8.13 SPICE-measured characteristics of the circuit in figure 8.11: (a) frequency response; (b) step response; (c) output (V_o) and equivalent input (V_i) noise

```
.PRINT   TRAN   V(13)
.PRINT   NOISE   INOISE   ONOISE
*INCLUDE   NONLIN.LIB
.END
```

Note the reference to the library NONLIN.LIB which contains the UA741 sub-circuit.

The step response, frequency response, and input and output noise are shown in figure 8.13. Note that the output noise spectral density tends to a constant value at high frequencies. This is determined by the noise from X2 since C_3 effectively provides a short-circuit at the input of the amplifier at these frequencies. The noise level in the stop band can be important. For example, if the circuit was to be used as an anti-aliasing filter prior to signal sampling, noise in the stopband above the Nyquist frequency would appear, together with the sampled signal.

References

Andrian, J. H., Yen, K. K. and Roig, G. (1987). 'Spice-compatible high frequency noise model of GaAs FET', in *Proceedings of IEEE 1987 Southeastcon 1*, pp. 191–193.

Hartmann, K. and Strutt, M. J. O. (1974). 'Computer simulation of small signal and noise behaviour of microwave bipolar transistors up to 12GHz', *IEEE Transactions on Microwave Theory and Techniques*, MT-22(3), 178–183.

Meares, L. W. and Hymowitz, C. E. (1988). *Simulating with Spice*, Intusoft, San Pedro, California.

Motchenbacher, C. D. and Fitchen, F. C. (1973). *Low-Noise Electronic Design*, Wiley, New York.

Rashid, M. H. (1990). *SPICE for Circuits and Electronics Using PSpice*, Prentice-Hall, Englewood Cliffs, New Jersey.

Scott, G. J. and Chen, T. M. (1987). 'Addition of excess noise in SPICE circuit simulations', in *Proceedings of IEEE 1987 Southeastcon 1*, pp. 186–190.

Skrzypkowiak, S. S. and Chen, T. M. (1987). 'A method of including resistor noise in SPICE circuit analysis', in *Proceedings of IEEE 1987 Southeastcon 1*, pp. 197–201.

Tuinenga, P. W. (1988). *SPICE: A Guide to Circuit Simulation and Analysis Using PSpice*, Prentice-Hall, Englewood Cliffs, New Jersey.

Whalen, J. J. and Paludi, C. (1977). 'Computer-aided analysis of electronic circuits – the need to include parasitic elements', *International Journal of Electronics*, 43(5), 501–511.

9   Low Noise Design

9.1 Introduction

The design specification of electronic equipment dealing with low-level signals should include limits on noise, signal-to-noise ratio, or on the functional reliability or error rate determined by these quantities. Failure to set noise or noise-determined limits, or to design to meet other specifications, leaving the 'noise problem' to be sorted out at the end, can lead to inelegant patching-up or expensive re-design. We are fortunate that, although noise affects the functioning of complete instruments, circuits susceptible to both intrinsic noise and layout-related noise (interference) are limited to low signal level circuits usually at the 'front-end' of the device, and it is possible to focus attention in this area. However, it should not be forgotten that interference emanates from other circuits and that actions to keep emitted or conducted noise to a minimum may be more cost-effective than actions taken at the susceptible circuit.

9.2 Layout/construction (electromagnetic compatibility)

9.2.1 Introduction

The noise in this category is characterised by a source, a path and a susceptible circuit. The noise in the susceptible circuit may be diminished by reduction at the source, by attenuation in the path and by reducing the susceptibility of the low-level circuit.

When the source is under the designer's control, it may be easiest to reduce the noise at source, by reducing voltage and current rise-times, or the size of radiating circuits, for example. It may also be easy to reduce radiation by shielding the source. The signal source and connecting cable may be shielded from field-coupled noise, but it is sometimes difficult to provide adequate shielding without interfering with function. An extreme example is an antenna source, which may not be shielded without rendering it ineffective as an antenna!

Many sources of noise are not under the designer's control – for example, radio transmitters and the consequences of miscellaneous electrical activity in radiated and power-line noise – and it is important that the device under design performs properly at the noise level of its working environment, which is not necessarily the same as in the prototype design laboratory. The noise-reducing design strategy must take this into account. It should also be remembered that intrinsic noise will remain, even when layout-related noise has been reduced to negligible levels, and there is little point in reducing the latter to much below the intrinsic noise level.

9.2.2 Preamplifier

One of the most important principles of low noise design is that the size of the susceptible circuit should be kept to a minimum to reduce field-coupled noise. To achieve this, the first amplifier in the signal or conditioning chain – the preamplifier – should be placed as near to the source as possible, and the preamplifier should have sufficient gain such that the total noise from all sources entering the signal chain later may be ignored. Comparing a source connected by a cable to an instrument and a similar device with a preamplifier between source and cable (figure 9.1) makes clear that, in the latter case, the noise introduced into the system by means of the cable – for example, field-coupled noise and movement-induced triboelectric and microphony noise – and the system noise performance degradation resulting from attenuation in the cable, may be rendered insignificant.

9.2.3 Shielding

Introduction

The purpose of shielding (screening) is to reduce electric, magnetic or electromagnetic field strengths by the use of conducting and high permeability containers. The shielding may be used to reduce the strength of fields emitted from a source by surrounding the source by shielding, or to reduce the strength of fields in the vicinity of a susceptible circuit by containing the circuit within a shield.

Calculation of the effectiveness of a shield is, in general, extremely complex. The electric and magnetic fields within a shielding container depend on the amplitudes and frequencies of the external fields (bearing in mind that these fields are affected by its presence), the relative orientation of container and fields, the container's size and shape, the shield material, the presence of seams and apertures, leads entering the container, its contents, and the position, within the container, at which the fields are measured.

In order to arrive at an understanding of the importance of these various factors and to give the practising engineer some formulae and figures on which to base designs, it has been necessary to introduce fairly gross simplifications into the analysis of shields.

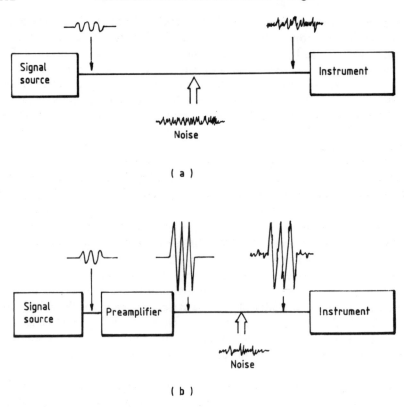

Figure 9.1   Signal source connected by cable to instrument: (a) without preamplifier; (b) with preamplifier

Probably the most common method of analysis of solid shields, and the one examined in most detail here, is the field impedance approach, introduced by Schelkunoff (1943) and developed by others (Vasaka, 1954; Schulz, 1971; Schulz *et al.*, 1988). In this method, the shield is assumed plane and of infinite lateral extent with the electric and magnetic field vectors parallel to the shield. The problem is thus reduced to a one-dimensional situation, allowing the use of relatively simple mathematics and giving an insight into the effect of shield and field characteristics, but leading to errors when used, unmodified, to calculate the fields inside real shielding containers.

Fields and impedance

Any alternating current has associated with it alternating magnetic and electric fields. The relative strengths of these fields depend on the nature of the circuit in which the current is flowing and on the position of the observation point relative to this source circuit. The ratio of the electric and magnetic fields is called the wave impedance. That is:

$$E/H = Z_w \tag{9.1}$$

If the source is primarily an electric field generator (no closed circuit) then the field close to this source is primarily electric and Z_w is high. If the source is primarily a magnetic field generator (closed circuit), then the magnetic field is dominant close to the source and Z_w is low. In the case of a small source, as the distance (r) from the source increases, the field which is dominant close to the source decreases as $1/r^3$ and the other field as $1/r^2$. Therefore the impedance of the electromagnetic field from an electric field source falls and that of the electromagnetic field of a magnetic field source rises. At large distances from the source, the field takes on the form of a plane electromagnetic wave with perpendicular E and H field vectors, both perpendicular to the direction of propagation (away from the source), with both field components reducing as $1/r$ and having a constant wave impedance which, in air, is:

$$Z_w = Z_0 = (\mu_0/\varepsilon_0)^{1/2} \tag{9.2}$$

$$= 377\,\Omega$$

The field impedances for the small electric and magnetic dipole sources of figure 3.1 are shown in figure 9.2, and the fields are depicted diagrammatically in figure 9.3 (see appendix D).

The transition between predominantly magnetic or electric field (near field region) and a constant Z_w-plane, progressive electromagnetic wave (far field region) takes place around $r = \lambda/2\pi$ in this case. Note that this is only true for the small sources considered. Transmitter antennae usually have dimensions comparable with λ and the near field is much longer.

The plane wave represents an energy loss to the radiating source and accounts for the radiation resistance of the source. The fields dominant close to each source may be calculated using the formulae of electrostatics and DC electromagnetism. They are fields in which energy is stored capacitatively (electric field) or inductively (magnetic field) and are called reactive fields. The impedance of the electric source near field in air is approximately:

$$Z_e \simeq (j2\pi f \varepsilon_0 r)^{-1} \tag{9.3}$$

and of the magnetic source near field:

$$Z_m \simeq j2\pi f \mu_0 r \tag{9.4}$$

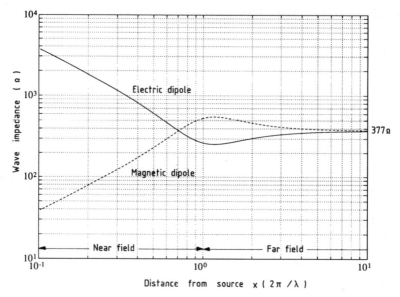

Figure 9.2   Wave impedance versus distance from source normalised to
$\lambda/(2\pi)$, for electric and magnetic dipole sources

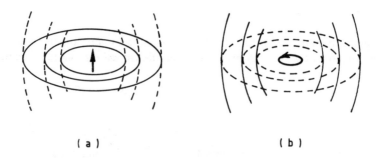

(a)   (b)

Figure 9.3   Radiated fields from small dipoles, H field, E field:
(a) electric dipole; (b) magnetic dipole

In general, the impedance of a plane wave in a medium is:

$$Z_w = (j2\pi f\mu/(\sigma + j2\pi f\varepsilon))^{1/2} \tag{9.5}$$

where σ is the conductance of the medium, and is called the characteristic
impedance of the medium.

So, we have for an insulator ($\sigma \ll 2\pi f\varepsilon$):

$$Z_i = (\mu/\varepsilon) \tag{9.6}$$

which becomes (9.2) in free space, and for a conductor ($\sigma \gg 2\pi f\varepsilon$):

$$Z_s = (j2\pi f\mu/\sigma)^{1/2} = (\pi f\mu/\sigma)^{1/2}(1+j) \tag{9.7}$$

or:

$$|Z_s| = (2\pi f\mu/\sigma)^{1/2}$$

$$= 3.68 \times 10^{-7}(\mu_r f/\sigma_r)^{1/2} \quad \Omega \tag{9.8}$$

where μ_r and σ_r are the permeability relative to free space and the conductance relative to copper respectively. That is:

$$\mu = \mu_r\mu_0$$

and:

$$\sigma = \sigma_r\sigma_c$$

where:

$$\sigma_c = 5.82 \times 10^7 \quad S\,m^{-1} \tag{9.9}$$

The impedance of good conductors is very small compared with the free space impedance. For example, the impedance of copper at 1 MHz is $3.68 \times 10^{-4}\,\Omega$.

Absorption

An electromagnetic wave in a medium is attenuated as a result of resistive losses (absorption). It is also known as penetration loss. The attenuation of a plane wave, as a result of absorption, follows the normal exponential law. That is:

$$E_x = E_0 \exp(-x/\delta)$$

and:

$$H_x = H_0 \exp(-x/\delta) \tag{9.10}$$

where E_x and H_x are the fields at depth x, and δ(skin depth) the depth at which the field has been reduced to $1/e$ of its initial ($x=0$) value is:

$$\delta = (\pi f\mu\sigma)^{-1/2} \tag{9.11}$$

For example, the skin depth for copper is 2.09 mm at 1 kHz and 0.0209 mm at 10 MHz.

The attenuation due to absorption in a shield of thickness x_s, is, from equations (9.10) and (9.11)):

$$A = -20\log_{10}(E_{xs}/E_0)$$

$$= -20\log_{10}(H_{xs}/H_0)$$

$$= 8.69(x_s/\delta) \quad \text{dB} \tag{9.12}$$

$$= 132x_s(f\mu_r\sigma_r)^{1/2} \quad \text{dB} \tag{9.13}$$

Reflection

Shielding depends, in addition to absorption, on reflection at impedance changes.

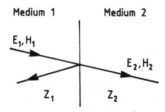

Figure 9.4   Reflection and transmission of wave at boundary

When a plane wave travelling in one medium is incident upon the plane boundary of a second medium, some of the wave is reflected and some transmitted (figure 9.4). In the case of perpendicular incidence, the electric field transmission coefficient is:

$$E_2/E_1 = 2Z_2/(Z_1 + Z_2) \tag{9.14}$$

and that of the magnetic field is:

$$H_2/H_1 = 2Z_1/(Z_1 + Z_2) \tag{9.15}$$

If medium 1 is an insulator (e.g. air) and medium 2 is a conductor, then $|Z_2| \ll |Z_1|$ and these equations may be written:

$$|E_2|/|E_1| = 2|Z_2|/|Z_1| \tag{9.16}$$

$$|H_2|/|H_1| = 2 \tag{9.17}$$

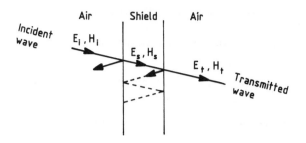

Figure 9.5 Transmission through shield

In the absence of absorption and multiple reflections within the shield, both of which may be considered separately, the reflection losses may be considered by reference to figure 9.5 which shows that for shields of impedance Z_s in air (making use of equations (9.14)–(9.17)):

$$E_t/E_i = (E_s/E_i)(E_t/E_s)$$

$$= 4Z_sZ_w/(Z_s + Z_w)^2$$

and:

$$H_t/H_i = 4Z_sZ_w/(Z_s + Z_w)^2 \tag{9.18}$$

and, for $|Z_s| << |Z_w|$:

$$|E_t|/|E_i| = |H_t|/|H_i| = 4|Z_s|/|Z_w| \tag{9.19}$$

Note that, although the overall reduction for both E and H fields is the same, the high reflection (low transmission) interface is the air–shield interface for the E field and the shield–air interface for the H field. This means that very thin screens may be used for E field shielding.

On the other hand, we may wish to have a greater thickness to make use of absorption in the case of magnetic fields.

Shield effectiveness

For all fields the reduction in field strength resulting from shielding (the shielding effectiveness) is (Vasaka, 1954; Schulz *et al.*, 1988):

$$S = R + A + B_s \quad \text{dB} \tag{9.20}$$

where R and A are the transmission coefficient and absorption in dB, and B_s is a correction term arising from multiple reflection within the shield. If there is sufficient absorption in the shield, this term is negligible.

The absorption term A is given by (9.12) or (9.13):

$$R = -20\log_{10}(|E_t|/|E_i|) \qquad \text{for electric fields}$$

$$= -20\log_{10}(|H_t|/|H_i|) \qquad \text{for magnetic fields}$$

$$= -20\log_{10}(4|Z_s|/|Z_w|) \qquad \text{for both fields if } |Z_s| << |Z_w| \qquad (9.21)$$

and:

$$B_s = 20\log_{10}|1 - ((Z_w - Z_s)/(Z_w + Z_s))^2 \exp(-2(1+j)x/\delta)| \qquad (9.22)$$

Except for the case of very low impedance fields, we have $|Z_s| << |Z_w|$, and the above may be simplified to:

$$B_s = 10\log_{10}(1 - 2 \times 10^{-0.1A}\cos(0.23A) + 10^{-0.2A}) \qquad (9.23)$$

This may be neglected for absorptions greater than 15 dB.
In all cases, $|Z_s|$ is given by (9.8).
For a far field plane wave, $Z_w = 377\,\Omega$ and:

$$R = 168 + 10\log_{10}(\sigma_r/(\mu_r f)) \quad \text{dB} \qquad (9.24)$$

For a near electric field Z_w is given by (9.3) and:

$$R = 322 + 10\log_{10}(\sigma_r/(\mu_r f^3 r^2)) \quad \text{dB} \qquad (9.25)$$

For a near magnetic field, Z_w is given by (9.4) and:

$$R = 14.6 + 10\log_{10}(\sigma_r f r^2/\mu_r) \quad \text{dB} \qquad (9.26)$$

where r is the distance of the shield from the source.
In general, absorption and reflection loss increase with conductivity. Increasing permeability increases absorption but decreases reflection loss. An increase in frequency leads to an increase in absorption and an increase in the reflection loss of magnetic fields, but a decrease in the reflection loss for plane wave and electric fields if the permeability is constant. For non-magnetic materials, such as copper and aluminium which are commonly used for shielding, the relative permeability is constant (unity). The permeability of magnetic materials, however, decreases with frequency and that of materials with high low-frequency permeability starts decreasing at lower frequencies. Different types of steel which again may be used for screen purposes have different permeabilities, but a typical silicon steel

has a low frequency relative permeability of 1000 falling to 500 at 10 MHz, whereas the relative permeability of μ-metal, which has a typical low frequency relative permeability of 2×10^4, falls to half this value at a few kHz.

The reflection losses for copper and steel ($\sigma_r = 0.1$) shields for the three field types at 10 Hz, 10 kHz and 10 MHz, at a distance of 0.1 metre from the source in the cases of the electric and magnetic near fields, are shown in table 9.1.

The absorption in 1 mm of copper and 1 mm of steel at these frequencies is shown in table 9.2.

Table 9.1   Reflection loss in dB

| Field | Frequency (Hz) | | |
	10	10^4	10^7
(a) Copper, $\sigma_r = \mu_r = 1$			
Electric	312	222	132
Magnetic	—	35	65
Plane wave	158	128	98
(b) Steel, $\sigma_r = 0.1$, $\mu_r = 10^3$ at 10 Hz and 10^4 Hz, and 0.5×10^3 at 10 MHz			
Electric	272	182	95
Magnetic	—	—	28
Plane wave	118	88	61

* Shield 10 cm from source for E and H fields
— indicates approximate formula invalid, small loss

Table 9.2   Absorption (dB) in 1 mm of shield

| Shield | Frequency (Hz) | | |
	10	10^4	10^7
Copper	0.4	13.2	417
Steel	4.2	132.0	2950
μ-metal	10.2		

It is clear that the electric field is easy to shield, using any good conductor, at any frequency. It is generally not necessary to rely on absorption, and thin shielding may be used. Plane waves also exhibit high reflection loss which may require augmentation by absorption loss at high radio frequencies, but again with thin shielding since the skin depth is very small at these frequencies.

As a result of the high reflection loss exhibited by plane waves and electric fields, useful shielding may be achieved with the use of very thin metallic films or foils (Liao, 1975; Hansen and Pawlewicz, 1982). In this case, absorption is negligible and the multiple reflection term B_s must be taken into account. If the shield is thin compared with the skin depth, and the sheet resistance ($R_s = (\sigma x_s)^{-1}$) is small compared with Z_w then the shield effectiveness of the film is:

$$S \simeq 20 \log_{10} |1 + Z_w/(2R_s)| \tag{9.27}$$

$$\simeq 20 \log_{10} |1 + 0.5 Z_w \sigma x_s| \tag{9.28}$$

Note that in the case of very thin films, when the film thickness is comparable with the electron mean-free-path in the film material, the effective conductivity (σ in equation (9.28)) is less than the bulk conductivity. Equation (9.27) is valid if R_s is the measured sheet resistance or is calculated using the effective conductivity for that sheet thickness. If the shield is a metallic film on a dielectric substrate, then the substrate thickness may have a significant effect on the shielding effectiveness (Klein, 1990).

A thin metallic film deposited on a transparent substrate can be used when it is required to combine good electromagnetic shielding at microwave frequencies and below with good transmission at optical frequencies – to view a display, for example. The reduced effective conductivity at optical frequencies means that light transmission is greater than at the lower frequencies.

Generally speaking, the low impedances of magnetic fields mean that it is necessary to rely on absorption to shield these fields. Above a few MHz, we may usually rely on absorption in non-magnetic conductors. At lower frequencies, it is necessary to use increasing thickness of magnetic material as the frequency falls, taking care that the material still has adequate permeability at the frequency of interest. In addition to absorption, concentration of magnetic flux (ducting) within high permeability shields reduces field strength and therefore increases shielding effectiveness. This effect is also covered in the next section.

At very low frequencies, a high permeability material such as μ-metal is required. As pointed out in chapter 3, section 3.2.3, these high permeability materials saturate at fairly low magnetic field strengths and it may be necessary to use a double shield arrangement. The permeability of these materials also falls if they are machined or otherwise mechanically or thermally stressed.

Circuit approach to shielding

At low frequencies, when the dimensions of the shielding container are small compared with the wavelength, a circuit approach (Bridges, 1988) can be taken to shielding effectiveness. In a low frequency electric field the induced charge distribution on the shielding container varies with the electric field. The associated current flow, in conjunction with the resistance of the container, leads to a voltage drop across the container and an electric field within. The current flow also leads to a magnetic field. Since charge on a conductor in an electric field concentrates at corners, so too do the current and internal fields. As the frequency decreases to zero, so too do the current flow and the internal fields, and the shield effectiveness becomes infinite, in agreement with the field impedance approach.

In a low frequency magnetic field, eddy currents are induced in the shielding container, giving rise to a magnetic field, in addition to the imposed field, and voltage drops leading to internal electric fields. The eddy currents concentrate at the edges of the container perpendicular to the imposed field, leading, again, to a non-uniformity of internal field. The shield eddy current is determined by the induced emf (proportional to the rate of change of the imposed magnetic field and therefore 90° out of phase with the field) and the shield impedance in the current path. At high frequency the shield impedance is mainly inductive, leading to a further phase shift of 90°, a magnetic field 180° out-of-phase with the imposed field and of similar magnitude – leading to almost complete cancellation of the imposed field and high shielding effectiveness. The degree of cancellation (and therefore, shielding effectiveness) depends on the shield resistance compared with its inductive reactance, and therefore increases with frequency. This is in qualitative agreement with the field impedance approach.

At zero frequency (DC), concentration (ducting) of the magnetic lines of force within a shield of magnetic material is the only means of shielding (Thomas, 1968), and remains a significant component of shielding effectiveness as the frequency increases until the effect of eddy currents becomes dominant. The shielding effectiveness resulting from ducting has been calculated for shielding containers of simple geometry. For a spherical container of radius a and thickness x_s $(<< a)$, made from a material of relative permeability μ_r $(> > 1)$, the shielding effectiveness is:

$$S_D = 20 \log_{10}\left(1 + \frac{2\mu_r x_s}{3a}\right) \tag{9.29}$$

At sufficiently high frequencies that the skin-depth is no longer small compared with the shield thickness, the skin effect leads to concentration of current flow on the outer surface of the container, thereby reducing currents and voltage drop on the internal surface and leading to a further increase in shielding effectiveness. This occurs for both electric and magnetic incident fields.

Equivalent circuits relating internal and external fields have been developed for simple shield geometries (Bridges, 1988).

The shielding effectiveness of a thin-walled conducting sphere for electric and magnetic fields, and for the cases of wall thickness (x_s) very much less, and very much greater, than the skin depth (low and high frequency respectively), are shown in table 9.3. The formula for the case of low frequency magnetic fields incorporates terms accounting for ducting when the shield material is magnetic (King, 1933). The magnitude of this effect, found by setting the frequency equal to zero, is consistent with (9.29) for high μ_r.

Table 9.3 Shielding effectiveness (dB) of a thin-walled conducting sphere (wall thickness $= x_s$, sphere radius $= a$)

Frequency	Electric field	Magnetic field
Low, $x_s \ll \delta$	$20 \log_{10}\left(\dfrac{\sigma x_s}{3\pi f \varepsilon_0 a}\right)$	$20 \log_{10}\left\|1 + 0.67\left(\dfrac{(\mu_r - 1)^2 x_s}{\mu_r a} + j\pi f \mu_0 \sigma a x_s\right)\right\|$
High, $x_s \gg \delta$	$20 \log_{10}\left(\dfrac{\sigma \delta \exp(x_s/\delta)}{6\sqrt{2\pi f \varepsilon_0 a}}\right)$	$20 \log_{10}\left(\dfrac{a \exp(x_s/\delta)}{3\sqrt{2}\mu_r \delta}\right)$

Apertures

Although the theory presented above suggests that very good shielding may be obtained with a suitable choice of material, it is only true of shields with no apertures. Since these are required for connections and ventilation they must be taken into account, and in practice they will usually determine the effectiveness of the shield. This is not to say that the preceding theory is of no account. It can be used to ensure that a solid shield gives shielding above that which is actually needed, but care should be taken to minimise field leakage through the necessary apertures in order to arrive at the shielding required.

Apertures have a greater effect on magnetic field shielding than that of electric fields, and therefore it is usually sufficient to consider only magnetic field leakage.

Since magnetic shielding essentially depends on field-induced eddy currents in the shield, not surprisingly, the degree of interruption of these currents determines the effect of holes in the shield. As a result of this, magnetic field leakage depends mainly on the largest dimension of an aperture, and a large number of small apertures having the same total area are preferable to one single hole (figure 9.6).

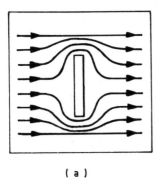 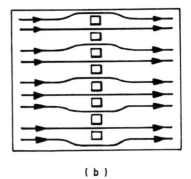

(a)   (b)

Figure 9.6   Effect of apertures on induced shield-current: (a) one large
aperture; (b) many small apertures

The effect of an aperture in a shielding container depends, in general, on
the nature of the incident field, the shield surface shape, aperture shape and
size, the region behind the aperture (container size, shape and contents), and
position and impedance of the susceptible circuit. As with the solid shield,
the problem has been simplified in order to arrive at simple but approximate
formulae.

Usually the aperture dimensions are small compared with the wavelength
of the incident field. In this case, except in the immediate vicinity of the
aperture (distance less than the aperture dimension), the field behind an
aperture in a thin shield has been shown to be approximately equal to the
near (inductive) fields of a number of magnetic and electric dipoles at the
position of the centre of the aperture (Bethe, 1944). For apertures having
two axes of symmetry (e.g. ellipse, rectangle) three are required – an electric
dipole perpendicular to the plane of the aperture, and two magnetic dipoles
within the plane and each parallel to an axis of symmetry. Their strengths
are determined by the field strengths either side of the shield in the absence
of the aperture (solid shield) (Butler *et al.*, 1978). In the case of a highly
reflecting and/or absorbing solid shield, only the incident fields need to be
considered. The dipole strengths are:

$$p = \alpha_e \varepsilon_0 E_p \qquad \text{(electric dipole)} \qquad (9.30)$$
$$m_a = \alpha_{ma} H_{ta} \qquad \text{(magnetic dipole, parallel to axis } a) \qquad (9.31)$$
$$m_b = \alpha_{mb} H_{tb} \qquad \text{(magnetic dipole, parallel to axis } b) \qquad (9.32)$$

where E_p is the electric field perpendicular to the shield and H_{ta} and H_{tb} are
the components of the tangential magnetic field H_t along the aperture axes
of symmetry, and α_e, α_{ma} and α_{mb} are called the polarisabilities of the
aperture. The polarisabilities have the dimensions of a volume and have

been determined for a number of simple shapes. They are shown for the circle and ellipse cases in table 9.4.

Table 9.4 Aperture polarisabilities from Cohn, 1951 and 1952; Quine, 1957; Butler *et al.*, 1978)

Shape	α_e	α_a	α_b
Circle radius a	$2a^3/3$	$4a^3/3$	$4a^3/3$
Thin ellipse, $a \gg b$ major semi-axis a minor semi-axis b	$\pi ab^2/3$	$\dfrac{\pi}{3}\dfrac{a^3}{\ln(4a/b)-1}$	$\pi ab^2/3$
Rectangle, length l width w $l=2w$ $l=5w$ $l=10w$	$0.0370l^3$ $0.0070l^3$ $0.0019l^3$	$0.1580l^3$ $0.0906l^3$ $0.0645l^3$	— — —

The fields at distance r from the dipoles, along a line perpendicular to the screen and through the aperture centre are:

$$E_s = \frac{p}{\pi\varepsilon_0 r^3} = \frac{\alpha_e E_p}{\pi r^3} \qquad (E \text{ field}) \qquad (9.33)$$

$$H_{sa} = \frac{m_a}{2\pi r^3} = \frac{\alpha_{ma}H_{ta}}{2\pi r^3} \qquad (\text{component of } H \text{ field parallel to } a) \qquad (9.34)$$

$$H_{sb} = \frac{m_b}{2\pi r^3} = \frac{\alpha_{mb}H_{tb}}{2\pi r^3} \qquad (\text{component of } H \text{ field parallel to } b) \qquad (9.35)$$

The variation of field strengths with position away from the perpendicular at constant r is not rapid. The fields on the shield at distance r from the aperture are of comparable magnitude.

In a highly reflecting shield, the fields E_p and H_t are double the incident fields E_{pi} and H_{ti} respectively (Taylor, 1973). Thus, for a circular aperture and plane waves at normal incidence ($E_p = 0$), the only significant field penetrating the aperture is the magnetic field and this is:

$$H_s = \frac{4a^3}{3\pi r^3}H_{ti} \qquad (9.36)$$

Electric field penetration will be important at grazing incidence, with the incident electric field orientated such that its component perpendicular to the shield is significant (Kraichman, 1966).

Note that for a thin ellipse (slot), the polarisability, and therefore the magnetic field penetration, are strongly dependent on the slot length but only slightly on its width if the field is parallel to the major axis. This illustrates the importance of ensuring good electrical contact in seams and around inspection hatches using (if necessary) wire mesh or conductive elastomer gaskets.

Note that there is some variation of notation in the literature dealing with this problem. The polarisabilities quoted in table 9.4 are those relating fields at the surface of the solid shield to the moments of dipoles located on a perfectly conducting shield. The fields (9.33)–(9.35) from such dipoles are double those from dipoles in free space (see appendix D). Some authors (Taylor, 1973; de Meulenaere and van Bladel, 1977) use a free-space dipole model, in which case the polarisabilities are double those shown in table 9.4. In addition, some authors (such as Kraichman, 1966) relate the free-space dipole moments to the incident rather than the surface fields, in which case the polarisabilities are doubled again and are four times those in table 9.4. Note also that some early papers use Gaussian rather than SI or MKS units.

If the shield thickness is comparable or greater than the aperture width, then the penetrating fields are smaller than calculated. This additional attenuation may be calculated approximately by treating the thick aperture as a waveguide operating below its cut-off frequency (Quine, 1957; Wheeler, 1964; McDonald, 1972).

The additional attenuation is dependent (although not strongly) on wave modes within the waveguide, which in turn are dependent on the incident wave orientation and polarisation with respect to the aperture axes. For a circular waveguide of diameter d and length l, and for plane waves at perpendicular incidence to the aperture, the appropriate mode is TE_{11} and the additional attenuation, for frequencies well below cut-off, is:

$$S = 32l/d \quad \text{dB} \tag{9.37}$$

The cut-off frequency for this mode in a circular cross-section waveguide of diameter d is (Terman, 1955):

$$f_c = 0.175 \times 10^9 / d \quad \text{Hz} \tag{9.38}$$

Equation (9.37) may be used for frequencies up to approximately $f_c/3$.

Note that this effect may be deliberately used to increase field attenuation by lengthening the aperture as shown in figure 9.7. Highly attenuating ventilation grills may be constructed using a honeycomb of waveguides.

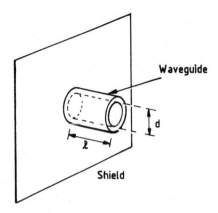

Figure 9.7 Waveguide aperture

Also note that field reductions by the aperture and waveguide effects are both independent of frequency within the constraints described.

We have assumed that there is no significant field penetration through the solid shield, and that the aperture is small compared with a wavelength. Since magentic field shielding for a non-magnetic shield increases as the frequency increases – from zero shielding at zero frequency – the magnetic field within a non-magnetic shielding container with apertures will decrease as the frequency increases until the field penetrating the solid shield is comparable with that penetrating the apertures. Well above the frequency at which these two components are equal, the internal field is dependent on aperture leakage and is approximately constant (Hoeft and Hofstra, 1988). There may be a trough in the plot of internal field versus frequency around the frequency at which the two components are equal if their phase difference leads to destructive interference.

At still higher frequencies, when the wavelength is comparable or greater than the aperture dimension, the simple dipole representation is not valid and interference between contributions from the points across the aperture lead to a strongly position-dependent field within the shield and resonant behaviour (Levy *et al.*, 1985).

The case of wire mesh and thin-film screens over an aperture may be treated in terms of a modification of the aperture polarisabilities (Casey, 1981). Note that at high frequencies when the reactance of the wire inductance is larger than its resistance, the shielding effectiveness of wire mesh falls with increasing frequency, and that the contact resistance between the mesh (or film) and the solid shield should be small compared with the mesh sheet-impedance in order to ensure that it does not impair shielding efficiency.

Circuit approach to apertures

The circuit approach may be used in connection with apertures. The voltage drop across an aperture resulting from shield surface currents induced by external fields will itself give rise to currents on the inner surface of the container. These inner surface currents will in turn give rise to internal fields. The voltage drop across the contact resistance in a seam will also give rise to currents on the internal surface of the shielding container. As with the penetration field described earlier, the worst case is when the external magnetic field is parallel to the seam, leading to a surface current perpendicular to the seam.

Magnetic flux leakage through an aperture induces a series emf in a wire within the shield. An electric field penetrating the aperture and terminating on a conductor gives rise to a current flow into the conductor. Bounds on the induced emf and current may be calculated by assuming that the conductor is linked by all the leakage flux (wire running across the aperture and perpendicular to the external field) and that the conductor intercepts all the penetrating electric field (Casey, 1981; Graf and Vance, 1988). In these cases, for a circular aperture of radius a, the maximum induced emf is:

$$V = j2\pi f \mu_0 a^2 H_t \tag{9.39}$$

and the current flowing into a conducting plate connected to the shield and placed across an aperture of area A such that it intercepts all the electric field flux penetrating the aperture is:

$$I = j2\pi f \varepsilon_0 A E_p \tag{9.40}$$

Note that these formulae are bounds and will usually greatly overestimate if misused.

9.2.4 Screened cables

Introduction

Part of the problem of shielding involves the connecting leads between devices. In order to be able to discuss shielding effectiveness, we first need to examine some properties of shielded cables.

Coaxial cable – screen-to-core coupling

The coaxial cable is probably the most common means of achieving screened connections. A short length of such a cable is shown in figure 9.8. The inner conductor and shield of this length of cable have a mutual inductance M_{is} and therefore constitute a transformer. Since all of the flux ϕ

associated with a current I_s in the shield encircles the inner conductor, if L_s is the inductance of the shield:

$$L_s = \phi/I_s \qquad (9.41)$$

and:

$$M_{is} = \phi/I_s \qquad (9.42)$$

and therefore:

$$M_{is} = L_s \qquad (9.43)$$

Note that we have assumed a uniform current density in a circular cross-section screen. Deviations from these conditions will lead to a small difference between M_{is} and L_s.

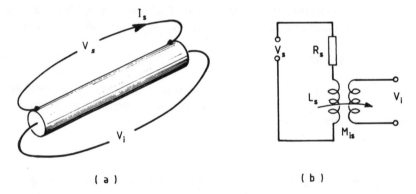

(a) $\qquad\qquad\qquad\qquad$ (b)

Figure 9.8 Coaxial cable as transformer: (a) cable; (b) equivalent circuit

It follows that the screen current induces an emf in the central conductor of:

$$V_i = j2\pi f M_{is} I_s \qquad (9.44)$$

or, making use of (9.43):

$$V_i = j2\pi f L_s I_s \qquad (9.45)$$

If R_s is the shield resistance and V_s is the voltage across the shield then this may be written:

$$V_i = (j2\pi f L_s V_s)/(R_s + j2\pi f L_s) \qquad (9.46)$$

or:

$$|V_i| = |V_s|/(1 + (f_c/f)^2)^{1/2} \qquad (9.47)$$

where:

$$f_c = R_s/(2\pi L_s) \qquad (9.48)$$

and is known as the shield cut-off frequency. A plot of $|V_i|/|V_s|$ against f_c/f is shown in figure 9.9. For frequencies very much greater than the shield cut-off frequency, $|V_i| = |V_s|$. A typical figure for f_c is 1 kHz.

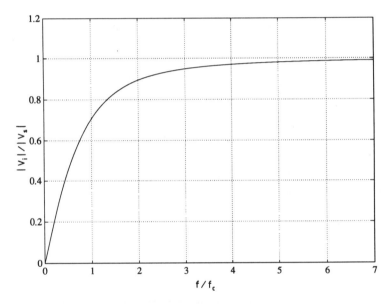

Figure 9.9   Ratio of core to screen voltage versus frequency normalised to shield cut-off frequency

Transfer impedance

One measure of shielding efficiency is transfer impedance (Vance, 1978). Consider the circuit in figure 9.10. The voltage appearing between the screen and inner conductor at one end of a length of coaxial cable when the other end is shorted and there is a screen current of I_s, is the sum of the voltage across the ends of the screen and the emf induced in the inner conductor resulting from the screen current and the mutual inductance between screen and inner conductor. Note that these voltages are of opposing signs. Then at low frequencies, such that skin depth effects are negligible:

$$V = I_s(R_s + j2\pi f L_s) - j2\pi f M_{is} I_s$$

$$= I_s R_s \tag{9.49}$$

Since R_s is linearly dependent on cable length, we may introduce a measure related just to cable type by normalising with respect to length. The cable screen transfer impedance is defined as the voltage appearing between the inner and shield conductors per unit length of cable and per unit screen current. That is:

$$Z_t = I_s^{-1} dV/dl \tag{9.50}$$

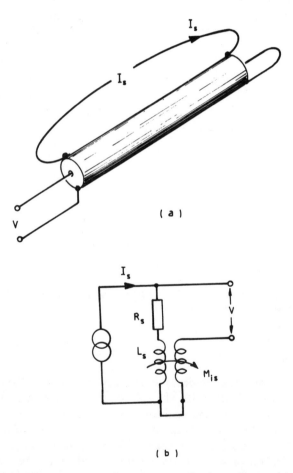

Figure 9.10 Measurement of screen transfer impedance: (a) cable connections; (b) equivalent circuit

In the case above:

$$Z_t = dR_s/dl \tag{9.51}$$

the resistance per unit length.

As the frequency increases, the skin-depth falls and, when it is comparable with the shield thickness, the transfer impedance starts to decrease. This is a result of the increasing concentration of shield current near the outer surface of the shield, the consequent decreasing fraction of screen current giving rise to a voltage drop on the inner surface. Typically, the skin depth is equal to the shield thickness at a few hundred kilohertz.

The above continues as the dominant process, as the frequency rises, for solid shield cables. However, in cables with non-solid screens (braid, for example) the effect of leakage of the flux, resulting from the outer shield current, into the cable leads to a difference between M_{is} and L_s, leading in turn to an increase in Z_t at higher frequencies. The frequency at which this effect becomes important depends on screen structure, and in particular, the percentage braid coverage in braid-screen cables. Typically this occurs in the frequency range of 500 kHz to 2 MHz (Vance, 1978). A plot of $|Z_t|/R_s$ (where R_s is the low frequency shield resistance in this case) against frequency for a typical cable is shown in figure 9.11.

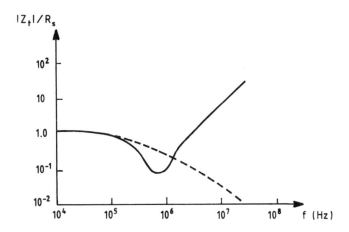

Figure 9.11 Screen transfer impedance normalised to low frequency screen resistance versus frequency for typical braided-screen cable. Dashed line – solid screen

Transfer admittance

The effectiveness of shielding against electric fields is best indicated by another figure of merit – transfer admittance – which is a measure of electric field leakage through the screen, or the mutual capacitance between the inner conductor and an external conductor (Vance, 1978). The transfer admittance is defined by:

$$Y_t = -V_s^{-1} dI_{sc}/dl \qquad (9.52)$$

where I_{sc} is the short-circuit current flowing between inner conductor and shield, per unit length of cable, as a result of a potential V_s between the cable screen and external conductor. The equivalent circuit is shown in figure 9.12. The current flows through the distributed mutual capacitance lC_{gc} between the external conductor and the cable core. This capacitance is very much smaller than the capacitance lC_{gs} between external conductor and screen, and through which most of the current from source V_s flows. Conductor impedances are assumed negligible compared with the reactances of the capacitances shown.

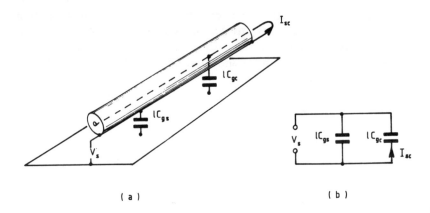

(a)

(b)

Figure 9.12   Measurement of transfer admittance

From the diagram and (9.52):

$$Y_t = j2\pi f C_{gc} \qquad (9.53)$$

Note that C_{gs} may be expressed as:

$$C_{gc} = kC_{gs}C_{sc} \qquad (9.54)$$

where C_{sc} is the capacitance per unit length between screen and core, and k is a function of screen structure and insulator permittivity. A typical value for k is 2×10^7 (Vance, 1978).

This leakage will normally only be important in high-impedance circuits since the small I_{sc} will give rise to significant voltage drop between core and cable only if there is a high impedance between these conductors.

Magnetic field screening

The near equality of M_{is} and L_s leads to an effect which is of use in screening magnetic fields using non-magnetic screening. Consider the circuit in figure 9.13, where a screened lead AB within circuit 2 (ABCD) runs close to a noise current (I_1) carrying lead in circuit 1. In the absence of the screen, the mutual inductance of the circuits is M_{21} and the emf generated in circuit 2 is:

$$V_2 = j2\pi f M_{21} I_1 \tag{9.55}$$

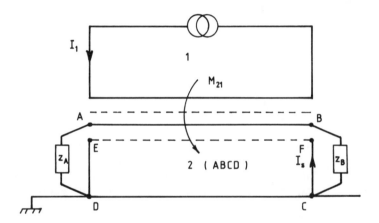

Figure 9.13   Magnetic field shielding using shield grounded at both ends

So long as no screen current flows – that is, the screen is unconnected, or connected to ground at only one point – then the magnetic flux in circuit 2 is unchanged and the same emf is induced. However, if the screen is connected to ground at both ends, since the mutual inductance of circuit 1 and the screen is equal to that of circuit 1 and the core AB (mutual inductance is substantially independent of conductor diameter in this situation), the emf induced in the screen circuit EFCD is also given by (9.55), and the screen circuit current is:

$$I_s = V_2/(R_s + j2\pi f L_s) \tag{9.56}$$

and this generates an emf in AB of:

$$V_{is} = j2\pi f M_{is} I_s \tag{9.57}$$

of opposite polarity to V_2. The net emf in circuit 2 is then (making use of the equality $M_{is} = L_s$ and equations (9.55)–(9.57)):

$$|V_2'| = M_{21}I_1(R_s/L_s)/(1 + (f_c/f)^2)^{1/2} \tag{9.58}$$

The variation with frequency of this noise voltage for both shielded and unshielded cable is shown in figure 9.14. The shield, when grounded at both ends, provides a significant degree of shielding at frequencies very much greater than the shield cut-off frequency.

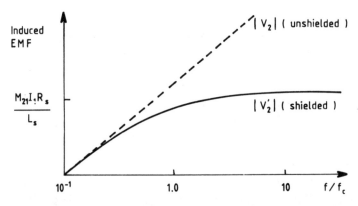

Figure 9.14 Induced emf in circuit ABCD of figure 9.13 for unshielded and shielded cable (shield grounded at both ends)

9.2.5 Low noise cabling and grounding

Introduction

The type, routing and ground connections of cables used to connect circuits and devices are very important considerations in low noise design. The types of noise most affected by these considerations are the field-coupled and the common path.

Twisted pair

The simplest method of reduction of field-coupled noise is the use of twisted pair connections as illustrated in figures 3.11(e) and 3.16(e). Note that it is essential that a floating source and differential amplifier are used to preserve a balance of coupling to both leads in the case of electric field (capacitative) coupling. If this is not possible, then a screened cable should be used.

With magnetic field noise, one end of one conductor may be grounded – but not both, since a balance of current flow is required in this case. Note that there is no rejection of electric field-coupled common mode noise, and an amplifier with an adequate common mode rejection ratio CMRR is required. This rejection also depends on a balance of impedance in the two leads. It is these requirements of balance and good CMRR which lead to problems at higher frequencies. As the frequency rises into the radio frequency range, amplifier-CMRR falls and stray reactance makes it difficult to achieve adequate balance.

Coaxial cable

This type of connection is much better for electric field shielding since, so long as there is a connection between the shield and ground, most of the noise current is simply shunted to ground. As we have seen in section 9.2.5, there is a mutual capacitance between the inner conductor and an outside conductor which will lead to some noise coupling, but it is small and can usually be reduced to negligible levels by the use of a cable with a high coverage screen.

A coaxial cable with only one screen connection to ground will give no magnetic field shielding. As shown in section 9.2.4, a ground connection at both ends will lead to some useful shielding well above the screen cut-off frequency. However, if there is significant noise voltage drop in the ground lead between the two connections (figure 9.15), this gives rise to a shield current and a noise voltage determined by this current and the shield transfer impedance (section 9.2.4). At frequencies above that at which the screen thickness is equal to the skin depth (typically a few hundred kHz) however, the noise current from this source concentrates at the outer surface of the shield as the frequency increases, and there is a progressive reduction of the mixing of ground noise and signal currents. This trend may, however, reverse at still higher frequencies (figure 9.11).

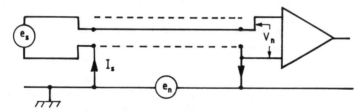

Figure 9.15 Ground connection noise voltage drop giving rise to shield current and noise voltage at amplifier input via shield transfer impedance

In the frequency range between shield and skin-depth cut-off frequencies, the decision to ground the screen at one or both ends depends on the relative contribution of magnetic field-coupled and ground-conductor noise.

The above assumes that the source and one amplifier input are grounded. If the source can float or a differential amplifier can be used, then the connections shown in figure 9.16 are preferred at frequencies below skin-depth cut-off, since they reduce magnetic field coupling by reducing the enclosed area and also avoid ground connection noise.

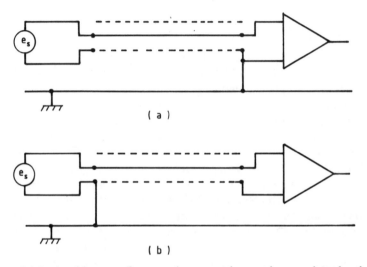

Figure 9.16   Avoidance of ground connection noise, and reduction of magnetic-field-coupled noise by enclosed area reduction: (a) floating source; (b) differential amplifier

At high frequencies, greater than the skin-depth cut-off, the advantage lies with double screen-end grounding, preferably with the screen close to a ground plane to reduce electromagnetic and magnetic field induced shield current, which in turn induces an emf in the inner conductor circuit by means of the cable screen transfer impedance. In fact, at high frequencies it is difficult to achieve isolation of the rest of the cable from ground if a single ground-plane connection is used, because of the capacitance between ground and screen. Note that a uniform rather than a pigtail connection of the screen to ground should be used (Paul, 1980) in order to avoid lack of screening of the cable ends and non-uniformity of screen current leading to an increased shield transfer impedance near the ends (figure 9.17).

If the frequency and/or cable length are sufficiently high that the length is a non-negligible fraction of a wavelength, multiple connections to ground

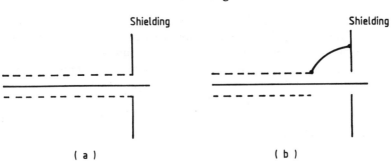

Figure 9.17 Cable termination. Uniform (a) cable screen termination preferred to (b) pigtail connection

should be used in order to avoid increased screen–core coupling as a result of a non-uniform screen potential.

If the cable is required to conduct signals with a wide frequency range, it is possible to achieve effective double-end grounding at high frequencies and single end grounding at low frequencies by making one ground connection a capacitor (figure 9.18) (Soldanels, 1967). A typical value is 3 nF which ideally has an impedance of approximately 5 Ω at 10 MHz and approximately 50 kΩ at 1 kHz. Note that minimising the parasitic self-inductance of such a capacitor requires very short leads, and the impedance at high frequencies may be higher than expected.

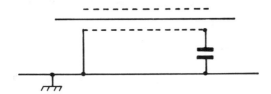

Figure 9.18 Effective single-end low frequency and double-end high frequency screen grounding

Screened twisted pair

The advantages of the twisted pair – particularly for reduction of magnetic field coupling – and the screened cable for reduction of electric field coupling may be combined in the screened twisted pair cable, and it is particulary useful at frequencies below skin-depth cut-off. Below this frequency, any signal on the screen will appear in the signal path if this path includes the screen, and the lead connections should avoid this. The preferred connections are shown in figure 9.19.

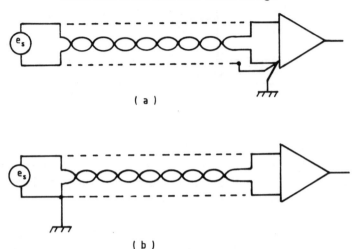

Figure 9.19   Preferred connections using screened twisted pair at low
              frequency: (a) floating source; (b) differential amplifier

9.2.6 *Breaking the ground loop*

There is often a need to connect two separately grounded circuits in such a
way that the noise potential difference between the two ground connections
does not appear in the signal circuit. A particular problem is the connection
between low-level analogue and digital circuits. Some methods are shown in
figure 9.20.

The transformer (figure 9.20(b)) is a simple option, particularly for a high
frequency analogue and pulse signals. At lower frequencies, the large
inductance required leads to bulky transformers and high interwinding
capacitance, leading to a degradation of the isolation of the two circuits.
Interwinding screens reduce this capacitance problem. If one screen is used,
it should be connected to the ground B and if there are two screens, then
they should be connected as shown. If a long connection is required, then
there may be some advantage in using two transformers, as shown in figure
9.20(c), in order to exploit the reduced noise-susceptibility of a balanced
(and perhaps screened) twisted pair.

This latter technique essentially gives circuit 1 a balanced output and
circuit 2 a differential input. If the circuits can be altered to these types, then
the ground voltage is reduced by the circuits' common mode rejection. The
common mode choke, as shown in figure 9.20(d), achieves the same effect by
allowing passage of differential signals unchanged while providing partial
cancellation of the common mode ground noise voltage. The common mode

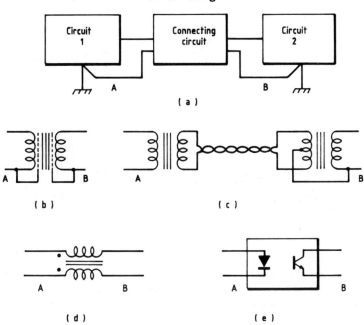

Figure 9.20 Breaking the ground loop: (a) block connnection diagram. Connecting circuit may be: (b) transformer; (c) two transformers and balanced line; (d) common mode choke; (e) opto-isolator

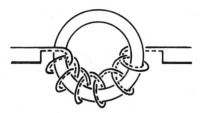

Figure 9.21 Common mode choke using ferrite core

choke is easily implemented at high frequencies by winding the pair of lines around a ferrite core (figure 9.21).

All the transformer techniques suffer from limited bandwidth and, if used at the signal source, may introduce significant noise if the loss resistance is appreciable.

The opto-isolator (figure 9.20(e)) provides a very high degree of isolation as the result of the low capacitance between input and output. Its linearity is poor, however, and it is usually used for pulse transmission.

9.2.7 Grounding

The major concern when considering ground connections, from a noise point of view, is the avoidance of coupling significant noise from noise-generating circuits into low-level signal circuits. This may be achieved by reducing noise currents flowing in ground connections, by the use of capacitors across the power-lines in noisy circuits – smoothing out the noisy variation in current demand and therefore ground connection current. The impedance of the common path may be reduced by the use of a low-impedance ground-plane (remembering that this is limited by skin-depth at the frequency of interest). Lastly, the common path may be eliminated.

At low frequencies, below about 1 MHz, the optimum solution is usually avoidance of common ground connections by using separate leads which are connected at one point as shown in figure 9.22(a). At higher frequencies, however, the increased inductive reactance of these leads gives rise to problems of circuit operation, radiated noise from the ground leads of the noise-generating circuits, and field-coupled noise in the ground cables of susceptible low-level circuits. In fact, these may also be considerations below 1 MHz.

When lead inductance becomes a problem, then the option of reducing the ground connection impedance should be taken and a common ground-plane used as shown in figure 9.22(b). Clearly, in order to derive the greatest benefit, the connections between circuits and ground-plane should be as short as possible.

If circuits are to operate over a wide frequency range then the hybrid grounding scheme shown in figure 9.22(c) may be used. This combines the advantage of ground-plane connection at high frequencies and single common point ground connection at low frequencies.

9.2.8 Filtering

In order to maintain the effectiveness of a shield it is important to reduce, to acceptable levels, the noise voltage difference between the shield and leads passing through the screen, and the noise current flowing in the conductor on the low noise side of the screen.

If the shielding encloses a susceptible circuit, this noise may be conducted from other circuits, or noise field-coupled to the lead, conducted into the shielded enclosure and re-radiated inside. The noise is reduced by filtering as

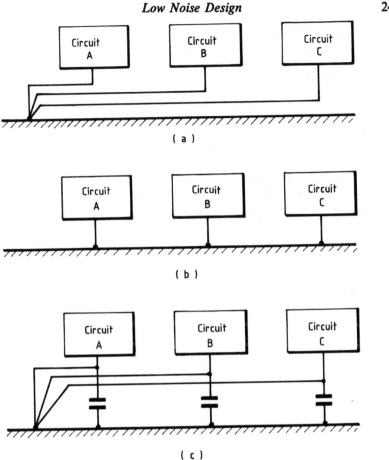

Figure 9.22 Grounding schemes: (a) single point; (b) multiple ground-plane connection; (c) hybrid

shown in figure 9.23. The shunt capacitor in power-line filtering may have to be limited in capacity since it increases the leakage current to ground, and this is limited for safety reasons. Power-line filters are usually purchased as complete devices and often contain transient suppressors in addition to filter components.

It is worth using a filter to limit signal path bandwidth at the point of entrance of the signal lead since this reduces the possibility of non-linearity demodulation problems (see section 3.2.2).

Coaxial feed-through capacitors are preferred in input filters since these have very low-inductance ground connections.

The same filtering arrangements can be used to reduce leakage of noise from a shielded noise-generating circuit.

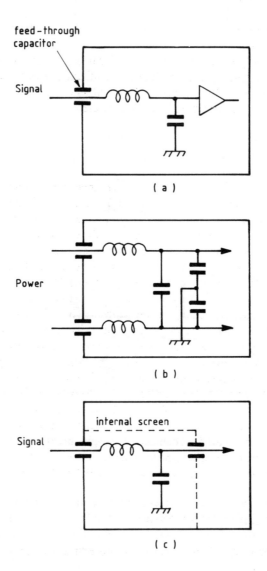

Figure 9.23   Filtering of leads entering a shielded enclosure: (a) signal;
            (b) power or balanced signal; (c) use of internal screen to
            reduce the radiation from filter

9.3 Intrinsic noise

9.3.1 Introduction

Low noise circuit design, as opposed to circuit layout, is concerned with the effect of intrinsic noise sources. This type of noise is more easily analysed and, if the noise is dominated by thermal and shot noise, circuit noise performance may be fairly accurately predicted.

9.3.2 Noise matching

In chapter 5, the optimum noise performance was discussed in terms of selecting, for an amplifier, a source resistance which gave the smallest fractional increase in noise over the unavoidable thermal noise of the source. Given a particular amplifier and source with a non-optimal resistance, the problem reduces to effectively changing the resistance that the source presents to the amplifier. It is essential to note that this should not be achieved by adding series or parallel resistance. These resistors contribute extra thermal noise and also attenuate the signal – leading to a smaller signal-to-noise ratio at the amplifier output. The solution to the apparent paradox – that the amplifier is, with added resistors, noise-matched and has a lower noise figure but the signal to noise ratio is worse – lies in the definition of noise figure (see equation (5.28)). The denominator in the defining equation is determined by the noise in the source – not in the source as modified by the addition of resistors. Another viewpoint is that the resistors are added to the amplifier, not the source, and degrade its noise performance. At the risk of labouring the point, two examples are considered in order to illustrate the problem. Consider an amplifier with noise figure F driven by a source with source resistance R_s which is not equal to the optimal source resistance R_{so} (figure 9.24(a)). Then, the noise figure is:

$$F = 1 + (e_N^2 + i_N^2 R_s^2)/(4kTR_s\Delta f) \qquad (9.59)$$

If R_s is less than R_{so}, then it is tempting to add a resistor R_{s1} in series with the source such that:

$$R_{so} = R_s + R_{s1} \qquad (9.60)$$

This leads to two extra noise sources in the input circuit – thermal noise (e_{t1}) in R_{s1} and a noise voltage drop (e_{i1}) across R_{s1} resulting from i_N (figure 9.24(b)). The noise figure is now:

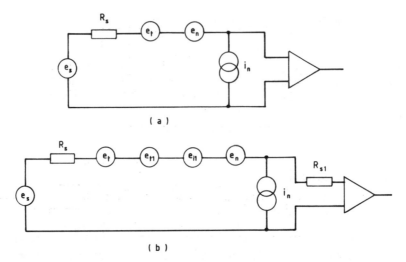

Figure 9.24 Change in noise performance when series resistance added: (a) original circuit; (b) with added series resistance R_{s1}

$$F' = 1 + (e_N^2 + i_N^2 R_s^2 + 4kTR_{s1}\Delta f + i_N^2 R_{s1}^2)/(4kTR_s\Delta f) \qquad (9.61)$$

$$= F + (4kTR_{s1}\Delta f + i_N^2 R_{s1}^2)/(4kTR_s\Delta f) \qquad (9.62)$$

If R_s is greater than R_{so} and a parallel resistor R_{s2} is added such that:

$$R_{so} = R_s//R_{s2} \qquad (9.63)$$

then again the amplifier sees an optimum source resistance. The circuit is shown in figure 9.25(a). The source e_{t2} is the thermal noise in R_{s2}. The analysis, in this case, is easier if the voltage generators are converted into equivalent current generators as shown in figure 9.25(b) (converting e_t, e_{t2} and e_s) and subsequently figure 9.25(c) (converting e_n). The noise figure is now:

$$F' = \left(i_T^2 + i_N^2 + i_{T2}^2 + e_N^2/(R_s//R_{s2})^2\right)\bar{i_T}^{-2} \qquad (9.64)$$

or, rearranging, and writing out i_T^2 and i_{T2}^2 specifically:

$$F' = F + R_s/R_{s2} + e_N^2(2 + R_s/R_{s2})(4kTR_{s2}\Delta f)^{-1} \qquad (9.65)$$

The noise figure has risen as a result of thermal noise in R_{s2} and an increased effect of amplifier noise as a result of attenuation of the source thermal noise (and signal).

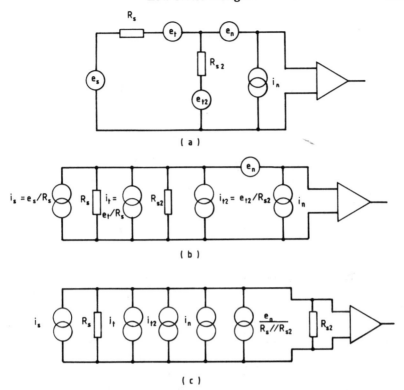

Figure 9.25 Change in noise performance when parallel resistance added: (a) circuit with added resistance R_{s2}; (b) signal and thermal noise voltage generators converted to current generators; (c) amplifier input voltage generator converted to equivalent current generator

In both cases there is an increase in noise figure – a degradation of noise performance.

Even added reactance may lead to an increase in noise figure. As shown in chapter 5, section 5.2.14, if the equivalent input noise generators are correlated, then the noise figure may be written (5.75):

$$F = F_{\min} + \frac{(G_s - G_{so})^2 + (B_s - B_{so})^2}{G_{nv}G_s} \tag{9.66}$$

Whatever the source conductance, the noise figure may be improved by adding a parallel reactive component to reduce $|B_s - B_{so}|$. The minimum noise figure is achieved if its susceptance is B_{so} (equal in magnitude and opposite in sign to the correlation susceptance). Conversely, if the added reactance increases $|B_s - B_{so}|$ then the noise figure will be worse. For example, at low frequencies, when the correlation susceptance is usually negligible, the addition of a reactive component, in parallel or in series, increases the noise figure even though it is not noise-generating. Series reactance increases the contribution of i_N to the equivalent input noise voltage as a result of the increased series impedance in the input circuit. Parallel reactance attenuates the source thermal noise (and signal) thereby increasing the relative contribution of the amplifier noise. At high radio frequencies, parasitic reactive components – for example, semiconductor lead inductance (Baechtold, 1971; Hartmann and Strutt, 1973; Anastassiou and Strutt, 1974; Gupta and Greiling, 1988) – should be taken into account when analysing noise performance.

The only way of improving the noise figure if we have a source with a non-optimal source resistance is to change the appearance of the source to the amplifier by the use of a transformer. The turns ratio is simply:

$$n = (R_{so}/R_s)^{1/2} \tag{9.67}$$

The equivalent source circuit seen by the amplifier is shown in figure 9.26. Since signal and source thermal noise are transformed by the same ratio then, if the transformer is ideal, the signal-to-noise ratio remains unchanged. That is, the noise factor of the transformer is unity. However, the parasitic resistive and reactive elements of real transformers will lead to a noise figure greater than optimum. In order to minimise this increase, the shunt and leakage inductances should be (respectively) very much greater and very much less than the source resistance in order to avoid source attenuation. The loss resistances (winding and core) are noise-producing and attenuating and these resistances, referred to the transformer primary, should be small compared with the source resistance.

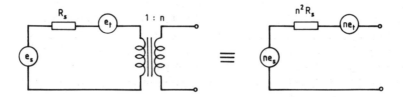

Figure 9.26   Transformer-coupled source

Note that the minimum noise figure depends on the product $\mathcal{E}_n \mathcal{I}_n$ (5.43). The minimum achievable when there is a choice of amplifier is that of the amplifier with the lowest $\mathcal{E}_n \mathcal{I}_n$ product used with a transformer to noise-match to the source.

9.3.3 Device selection

The preceding discussion presumes that both the amplifier noise characteristics and the source impedance are fixed, and the only variable is the method of connection. Although the theory of noise matching was developed by considering fixed amplifier characteristics and variable source impedance, in practice the designer normally has a choice of amplifiers and will need to choose one with an acceptable F_{min} in order to meet signal-to-noise ratio specification even if a transformer is used to alter the effective source impedance. Often, however, particularly at low frequencies, transformers with an acceptably high winding inductance are unacceptably large, and the designer is left with a source of fixed impedance and a choice of amplifiers. If $\mathcal{E}_n$ and $\mathcal{I}_n$ are the variables in equation (5.34), then there is no minimum and the noise figure decreases monotonically as $\mathcal{E}_n$ and $\mathcal{I}_n$ decrease. In this case it is necessary to select an amplifier such that the total noise is below a specification threshold, if this is set, or such that the equivalent input noise arising from $\mathcal{E}_n$ and $\mathcal{I}_n$, that is:

$$\mathcal{E}_a^2 = \mathcal{E}_n^2 + \mathcal{I}_n^2 |Z_s|^2 \tag{9.68}$$

where Z_s is the source impedance ($Z_s = R_s + jX_s$), is small compared with the source thermal noise:

$$\mathcal{E}_t^2 = 4kTR_s \tag{9.69}$$

when the lowest reasonable noise level is required.

There is normally little advantage to be gained in having an amplifier noise which is very much less than the source thermal noise, bearing in mind that an $\mathcal{E}_t/\mathcal{E}_n$ ratio of 3 leads to a total rms equivalent input noise voltage only 5 per cent greater than $\mathcal{E}_t$ (section 2.2.9). However it is important to remember that an antenna may have a noise temperature well below the standard temperature of 290 K, and an amplifier which is 'low noise' when operating with a source at this temperature may not be low noise when operating with sources of significantly lower equivalent temperature.

It follows from the above that a choice of amplifier depends strongly on the value of Z_s. If Z_s is high then it is important to keep $\mathcal{I}_n$ low, and if Z_s is low then a low value of $\mathcal{E}_n$ is required.

9.3.4 Choice of input stage

In a well designed amplifier, the noise performance is determined by the first stage and for this there is a choice of bipolar transistor or FET. The low $\mathcal{I}_n$ of FETs and in particular MOSFETs, together with $\mathcal{E}_n$ values comparable or slightly greater than that of bipolar transistors operating under low noise conditions, mean that these devices have a clear advantage when operating with high source impedances. On the other hand, bipolar transistors with low $r_{bb'}$ have an advantage at very low source impedances. However, there is no clear dividing line below which a bipolar transistor should be chosen and above which we should select an FET. With a source impedance of a few ohms or tens of ohms, a bipolar transistor with $r_{bb'}$ of the same order or preferably less than the source resistance and operating at a collector current in the range of a few tenths to a few milliamps is preferred, as may be seen from figure 6.9. High β low noise transistors with $r_{bb'}$ of less than $10\,\Omega$ are available and the opposing dependence of the shot noise components of $\mathcal{E}_n$ and $\mathcal{I}_n$ on collector current (figure 6.9) enables noise matching at a low F_{min}. The relatively high collector current also leads to a high f_T and thereby the maintenance of low noise performance at high frequencies. In the source resistance range of a few tens of ohms to about $10\,k\Omega$, either bipolar transistors or FETs may be used, although the lower collector current required for low noise operation with high source impedances leads to low f_T and therefore a limited frequency range for the bipolar devices. Since at high frequencies, power, as well as noise matching, is often important, the low input impedance of the common base configuration is often used.

For source impedances above about $10\,k\Omega$, the FET with its low $\mathcal{I}_n$ has a clear advantage and, since the advantage of the bipolar transistor becomes obvious only at very low source impedances, the FET is the device of choice when the input is to be driven from a range of source impedances.

At audio and sub-audio frequencies, the flicker-noise corner frequency is important and should be as low as possible. At these frequencies and with moderate or high source impedances, the bipolar transistor will be operating at low collector currents and flicker noise is apparent in $\mathcal{I}_n$ and not in $\mathcal{E}_n$. In FETs, flicker noise is apparent in $\mathcal{E}_n$ only. The corner frequency tends to be lower for bipolar transistors and there may be some advantage in considering these even for moderately high source impedances ($10\,k\Omega$ – $1\,M\Omega$) at low frequencies. However, it is possible to find high quality FETs with good low frequency, low noise performance. The noise voltage at $10\,Hz$ is often quoted rather than the corner frequency and devices with noise levels of around $1\,nV\,Hz^{-1/2}$ are available. MOSFETs have such a low current noise at low frequencies that it is usually not quoted, and the device is considered to have only voltage noise below the frequency at which the channel thermal noise capacitatively coupled to the gate becomes significant. They are, for this reason, very attractive for very high source impedance applications. However, the $\mathcal{E}_n$ flicker noise corner frequencies tend to be much higher, usually greater than $10\,kHz$, and this militates against their use even at audio frequencies.

Integrated circuit amplifiers tend to have poorer noise performance than discrete devices as a result of the compromises necessary in their design, and for the lowest achievable noise it is preferable to have a discrete device for the first stage. Nevertheless, low noise devices are available and may well meet the required specifications.

9.3.5 Feedback

Often we require not only noise matching to ensure low noise figures but also power (impedance) matching in order to maximise power transfer and/ or to avoid reflection at cable termination. Altering the impedance at the source by adding series or parallel impedance is not a useful method of achieving this since, as we have seen earlier, this reduces the signal-to-noise ratio. However, we can use negative feedback to alter the input impedance – shunt feedback to reduce the input impedance and series voltage feedback to increase it – and, as shown in chapter 5, section 5.2.11, provided the additional impedance in the source circuit is negligible compared with the source resistance and, for shunt feedback, the feedback impedance is high, then the noise characteristics of the amplifier are substantially unchanged. (This assumes that the system bandwidths are equal.) Provided that the open loop gain is sufficiently high, these constraints on the feedback components can usually be met.

Feedback, of course, may be used not only to achieve impedance matching but also to give high input impedance in order to reduce the shunting of the source or a low input impedance for current, rather than voltage, amplification.

9.3.6 Biasing

As we have seen, any extra impedance in the source circuit, other than reactance to cancel the correlation susceptance of the amplifier, leads to an increase in the noise figure. This poses a problem in considering how to bias the input stage. We illustrate the problems and solutions by considering a bipolar transistor, common emitter amplifier (figure 9.27(a)), but the principles are applicable to other devices.

In the emitter circuit, R_{e1} and R_{e2} are used for bias point stabilisation and the unbypassed resistor R_{e1} introduces some negative feedback. Both thermal and excess noise are generated in these resistors. If C_e is sufficiently large that R_{e2} is effectively bypassed at the lowest frequency of interest, then only the noise from R_{e1} need be considered. Since this noise is effectively in series with the source, it is important that the noise is small compared with the thermal noises in the source resistance R_s. This implies:

$$R_{e1} \ll R_s \qquad (9.70)$$

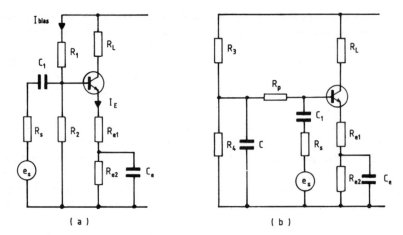

Figure 9.27 Biasing of common emitter stage: (a) normal circuit; (b) low noise design

and that it is a low noise resistor generating negligible excess noise compared with the thermal noise in R_s. The bypass capacitor C_e should have a sufficiently low reactance at the lowest frequency of interest to reduce the noise from R_{e2} to a negligible level. This may be a more severe constraint than that imposed by frequency response considerations.

The bias resistors R_1 and R_2 effectively appear in parallel across the amplifier input, and their thermal and excess noise appear in the input circuit. If R_p ($R_1//R_2$) is sufficiently large compared with R_s then their thermal and excess noise, attenuated by the voltage divider formed by R_p and R_s, may be small compared with the thermal noise in R_s. Clearly, R_1 and R_2 should have low excess noise.

Making R_1 and R_2 too large compromises the bias point stability. Usually I_{bias} is arranged to be about four times I_E. The bias point stability is, in fact, dependent on R_p since the change in base potential as a result of a bias current change ΔI_B is:

$$\Delta V_B = R_p \Delta I_B \tag{9.71}$$

An arrangement which gives essentially the same bias stability and same thermal noise contribution from the bias resistor but reduced excess noise is shown in figure 9.27(b) (note $R_3//R_4 \ll R_p$). The excess noise is reduced because only the base bias current flows through R_p, and noise from R_3 and R_4 is shunted by C. Since R_p shunts the source, low noise design requires that R_p also be large compared with R_s. A compromise between this requirement and that of adequate bias point stability may be necessary.

Frequency response considerations determine that C_1 be sufficiently large that its reactance is small compared with $(R_s + R_p//R_i)$(where R_i is the input impedance of the amplifier at the base – ground port) at the lowest frequency of interest. Consideration of the increasing reactance of C_1 at low frequencies, leading to an increased noise contribution from the noise voltage drop across C_1 arising from the equivalent input noise current generator, particularly if the noise current is dominated by flicker noise, may lead to a larger value for C_1 than dictated by frequency response considerations alone. The reactance (X_c) of C should be sufficiently small to reduce the thermal and excess noise from R_3 and R_4 to a negligible level compared with the source thermal noise. This may be a more severe constraint than $X_c << R_3//R_4$.

The avoidance of extra impedance in the input circuit on noise performance grounds means that, in general, bandwidth-determining components should be used at a later stage.

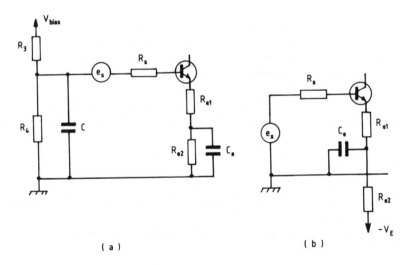

(a)　　　　　　　　　　　　(b)

Figure 9.28　Biasing via source: (a) floating source; (b) grounded source

If the source is able to tolerate the bias current flowing through it, the arrangements shown in figure 9.28, which do not require any unbypassed base bias resistors, may be used.

At radio frequencies, the bias resistor problem is avoided by the use of rf chokes or biasing through transformer or tuned circuit (figure 9.29).

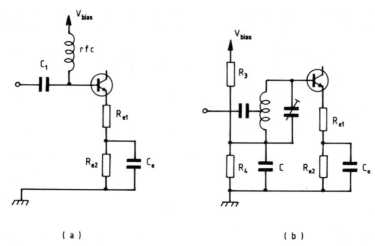

Figure 9.29   Biasing at radio frequencies: (a) using radio frequency choke (rfc); (b) using tuned circuit

References

Anastassiou, A. and Strutt, M. J. O. (1974). 'Effect of source lead inductance on the noise figure of a GaAs FET', *Proceedings of the IEEE*, **62**, 406–408.

Baechtold, W. (1971). 'Noise behavior of Schottky barrier gate field-effect transistors at microwave frequencies', *IEEE Transactions on Electron Devices*, **ED-18(2)**, 97–104.

Bethe, H. A. (1944). 'Theory of diffraction by small holes', *Physical Review*, **66**, 163–182.

Bridges, J. E. (1988). 'An update on the circuit approach to calculate shielding effectiveness', *IEEE Transactions on Electromagnetic Compatibility*, **30(3)**, 211–221.

Butler, C. M., Rahmat-Samii, Y. and Mittra, R. (1978). 'Electromagnetic penetration through apertures in conducting surfaces', *IEEE Transactions on Electromagnetic Compatibility*, **EMC-20(1)**, 82–93.

Casey, K. F. (1981). 'Low-frequency electromagnetic penetration of loaded apertures', *IEEE Transactions on Electromagnetic Compatibility*, **EMC-23(4)**, 367–377.

Cohn, S. B. (1951). 'Determination of aperture parameters by electrolytic-tank measurements', *Proceedings of the IRE*, 39, 1416–1421 (errata in *Proc. IRE*, **40** (1952) 33).

Cohn, S. B. (1952). 'The electric polarizability of apertures of arbitrary shape', *Proceedings of the IRE*, **40**, 1069–1071.

de Meulenaere, F. and van Bladel, J. (1977). 'Polarizability of some small apertures', *IEEE Transactions on Antennas and Propagation*, **AP-25(2)**, 198–205.

Graf, W. and Vance, E. F. (1988). 'Shielding effectiveness and electromagnetic protection', *IEEE Transactions on Electromagnetic Compatibility*, **30(3)**, 289–293.

Gupta, M. S. and Greiling, P. T. (1988). 'Microwave noise characterization of GaAs MESFETs: determination of extrinsic noise parameters', *IEEE Transactions on Microwave Theory and Techniques*, **36(4)**, 745–751.

Hansen, R. C. and Pawlewicz, W. T. (1982). 'Effective conductivity and microwave reflectivity of thin metallic films', *IEEE Transactions on Microwave Theory and Techniques*, **30(11)**, 2064–2066.

Hartmann, K. and Strutt, M. J. O. (1973). 'Changes of the four noise parameters due to general changes of linear two-port circuits', *IEEE Transactions on Electron Devices*, **ED-20(10)**, 874–877.

Hoeft, L. O. and Hofstra, J. S. (1988). 'Experimental and theoretical analysis of the magnetic field attenuation of enclosures', *IEEE Transactions on Electromagnetic Compatibility*, **30(3)**, 326–340.

King, L. V. (1933). 'Electromagnetic shielding at radio frequencies', *Philosophical Magazine and Journal of Science*, **15(97)**, 201–223.

Klein, C. A. (1990). 'Microwave shielding effectiveness of EC-coated dielectric slabs', *IEEE Transactions on Microwave Theory and Techniques*, **38(3)**, 321–324.

Kraichman, M. B. (1966). 'The near field beyond a perfectly conducting plane screen with a small circular aperture', *IEEE Transactions on Antennas and Propagation*, **AP-14**, 389–390.

Levy, P. H., Faulkner, J. E. and Shaeffer, D. L. (1985). 'Short-pulse microwave coupling to apertures in a conducting plane', *IEEE Transactions on Nuclear Science*, **NS-32(6)**, 4333–4339.

Liao, S. Y. (1975). 'Light transmittance and microwave attenuation of a gold-film coating on a plastic substrate', *IEEE Transactions on Microwave Theory and Techniques*, **MTT-23**, 846–849.

McDonald, N. A. (1972). 'Electric and magnetic coupling through small apertures in shield walls of any thickness', *IEEE Transactions on Microwave Theory and Techniques*, **MTT-20(10)**, 689–695.

Paul, C. R. (1980). 'Effect of pigtails on crosstalk to braided-shielded cables', *IEEE Transactions on EMC*, **EMC-22(3)**, 161–172.

Quine, J. P. (1957). 'Theoretical formulas for calculating the shielding effeectiveness of perforated sheets and wire mesh screens', in *Proceedings of the 3rd Conference on Radio Interference Reduction*, Armour Research Foundation, pp. 315–329.

Schelkunoff, S. A. (1943). '*Electromagnetic Waves*, van Nostrand, New York.

Schulz, R. B. (1971). 'Shielding', in *Practical Design for Electromagnetic Compatibility* (Ficchi, R. F., ed.), Hayden, New York, pp. 69–92.

Schulz, R. B., Plantz, V. C. and Brush, D. R. (1988). 'Shielding theory and practice', *IEEE Transactions on Electromagnetic Compatibility*, **30(3)**, 187–201.

Soldanels, R. M. (1967). 'Elimination of structure-current effects in RF return paths', *IEEE Transactions on Electromagnetic Compatibility*, **EMC-9(2)**, 23.

Taylor, C. D (1973). 'Electromagnetic pulse penetration through small apertures', *IEEE Transactions on Electromagnetic Compatiblity*, **EMC-15(1)**, 17–26.

Terman, F. E. (1955). *Electronic and Radio Engineering*, McGraw-Hill, New York.

Thomas, A. K. (1968). 'Magnetic shielded enclosure design in the DC and VLF region', *IEEE Transactions on Electromagnetic Compatibility*, **EMC-10(1)**, 142–152.

Vance, E. F. (1978). '*Coupling to Shielded Cables*, Wiley, New York.

Vasaka, C. S. (1954). 'Problems in shielding electronic equipment', in *Proceedings of the Conference on Radio Interference Reduction, Chicago, 1954*, Armour Research Foundation, pp. 86–103.

Wheeler, H. A. (1964). 'Coupling holes between resonant cavities or waveguides evaluated in terms of volume ratios', *IEEE Transactions on Microwave Theory and Techniques*, **MTT-12**, 231–244.

Appendix A: Constants

Table A.1 Physical constants

Quantity	Symbol	Value
Electronic charge	e	1.602×10^{-19} C
Speed of em radiation	c	2.998×10^{8} m s^{-1}
Permittivity of free space	ε_0	8.854×10^{-12} Fm^{-1}
Permeability of free space	μ_0	$4\pi \times 10^{-7}$ Hm^{-1}
Boltzmann's constant	k	1.380×10^{-23} JK^{-1}
Planck's constant	h	6.62×10^{-34} Js

Table A.2 Material constants, μ_r and σ

Material	μ_r	σ (Sm^{-1})
Aluminium	1.00	3.82×10^{7}
Copper	1.00	5.82×10^{7}
Silver	1.00	6.17×10^{7}
Gold	1.00	4.10×10^{7}
Carbon steel (typical)	1 000 (100 Hz)	0.6×10^{7}
μ-metal	20 000 (100 Hz)	0.2×10^{7}

Table A.3 Material constants, ε_r

Material	ε_r
Air	1.0006
Bakelite	4.74
Perspex	3.45
Polythene	2.26
Polystyrene	2.55

Appendix B:
Noise Model of Linear Two-port Network

Consider the linear two-port shown in figure B.1.

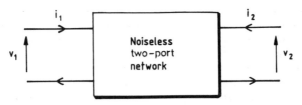

Figure B.1   Noiseless two-port network

If the network is lossless, then, in terms of the Fourier transforms of the voltages and currents at the two ports, linear network theory gives the relationships:

$$V_1 = Z_{11}I_1 + Z_{12}I_2$$

$$V_2 = Z_{21}I_1 + Z_{22}I_2 \tag{B.1}$$

where the impedance matrix $\begin{bmatrix} Z_{11} & Z_{12} \\ Z_{21} & Z_{22} \end{bmatrix}$ describes the input, output and transfer characteristics of the network.

Noise is generated in a real two-port network (figure B.2(a)) and, by considering each port in turn, Thévenin's theorem gives the equivalent circuit shown in figure B.2(b), where e_{n1} and e_{n2} are noise voltage generators, each with a spectral density equal to that measured at the respective port of the real network with the other port open-circuit.

Note that the noise generators are, in general, partially correlated since they derive from different fractions of the same internal noise sources.

The network equations now are:

$$V_1 + E_{n1} = Z_{11}I_1 + Z_{12}I_2$$

$$V_2 + E_{n2} = Z_{21}I_1 + Z_{22}I_2 \tag{B.2}$$

255

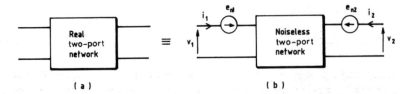

Figure B.2   (a) Real two-port network. (b) Thévenin equivalent

Consider, now, a two-port network with a series voltage and parallel current generator at port 1, as shown in figure B.3.

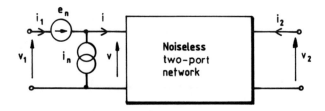

Figure B.3   Alternative equivalent circuit of real two-port network

Then:

$$V = Z_{11}I + Z_{12}I_2$$

$$V_2 = Z_{21}I + Z_{22}I_2 \qquad \text{(B.3)}$$

and:

$$I = I_1 + I_n$$

$$V = V_1 + E_n \qquad \text{(B.4)}$$

Combining (B.3) and (B.4) and rearranging gives:

$$V_1 + (E_n - Z_{11}I_n) = Z_{11}I_1 + Z_{12}I_2$$

$$V_2 - Z_{21}I_n = Z_{21}I_1 + Z_{22}I_2 \qquad \text{(B.5)}$$

If these equations are compared with (B.2) we see that the models of figures B.2(b) and B.3 are equivalent if:

$$E_{n1} = E_n - Z_{11}I_n$$

$$E_{n2} = -Z_{21}I_n \tag{B.6}$$

Since E_{n1} and E_{n2} are, in general, partially correlated, so too are E_n and I_n.

Note that a similar result can be derived by considering the admittance rather than the impedance matrix of the network, and by considering the Norton equivalent noise current generators at the two ports rather than the voltage generators considered here.

Appendix C: Noise Descriptors

Name	Symbol (units)	Comments/relationships
Voltage and current time-function	e_n (volt) i_n (amp)	Strictly $e_n(t)$, $i_n(t)$. Written with time-dependence implied
Mean square value	e_N^2 (volt^2) i_N^2 (amp^2)	$e_N^2 = \overline{e_n^2(t)}$ $i_N^2 = \overline{i_n^2(t)}$
Root mean square value	e_N (volt) i_N (amp)	$e_N = \sqrt{\overline{e_n^2(t)}}$ $i_N = \sqrt{\overline{i_n^2(t)}}$
Fourier transform	E_n (volt/Hz) I_n (amp/Hz)	Strictly $E_n(f)$, $I_n(f)$ Written with frequency dependence implied $$E_n(f) = \int_{-\tau/2}^{\tau/2} e_n(t)e^{-j2\pi ft}\,dt$$ $$I_n(f) = \int_{-\tau/2}^{\tau/2} i_n(t)e^{-j2\pi ft}\,dt$$ $(\tau \to \infty)$
Normalised spectral density	$S_n(f)$ or $S_{nv}(f)$ (volt^2/Hz) $S_{ni}(f)$ (amp^2/Hz)	$$S_n(f) = \lim_{\tau\to\infty}\left[\frac{2\overline{|E_n(f)|^2}}{\tau}\right]$$ $$S_{ni}(f) = \lim_{\tau\to\infty}\left[\frac{2\overline{|I_n(f)|^2}}{\tau}\right]$$ If e_N and i_N measured in narrow band Δf, then: $$S_n(f) = e_N^2/\Delta f$$ $$S_{ni}(f) = i_N^2/\Delta f$$

Name	*Symbol (units)*	*Comments/relationships*
Root normalised spectral density	$\mathcal{E}_n$ (volt/$\sqrt{Hz}$) $\mathcal{I}_n$ (amp/$\sqrt{Hz}$)	Strictly $\mathcal{E}_n(f)$ and $\mathcal{I}_n(f)$. Written with frequency-dependence implied $$\mathcal{E}_n = \sqrt{S_n(f)} = e_N/\sqrt{\Delta f}$$ $$\mathcal{I}_n = \sqrt{S_{ni}(f)} = i_N/\sqrt{\Delta f}$$ These are the most commonly used descriptors for noise
Available power spectral density	P_n (watt/Hz)	Strictly $P_n(f)$. Written with frequency-dependence implied $$P_n(f) = S_n(f)/(4R) = e_N^2/(4R\Delta f)$$ or: $$P_n(f) = S_{ni}(f)R/4 = i_N^2 R/(4\Delta f)$$

Appendix D: Dipole Fields

In spherical (r, θ, ϕ) coordinates, using standard notation, the components of the magnetic and electric fields of an electric dipole of oscillatory moment:

$$p = p_0 e^{j2\pi ft} \tag{D.1}$$

situated at the origin and directed along the polar axis, are:

$$E_r = \frac{2p\beta^3}{4\pi\varepsilon_0} \left(\frac{1}{\beta^3 r^3} + \frac{j}{\beta^2 r^2} \right) \cos(\theta) e^{-j\beta r}$$

$$E_\theta = \frac{p\beta^3}{4\pi\varepsilon_0} \left(\frac{1}{\beta^3 r^3} + \frac{j}{\beta^2 r^2} - \frac{1}{\beta r} \right) \sin(\theta) e^{-j\beta r}$$

$$E_\phi = 0$$

$$H_r = 0$$

$$H_\theta = 0$$

$$H_\phi = \frac{cp\beta^3}{4\pi} \left(\frac{j}{\beta^2 r^2} - \frac{1}{\beta r} \right) \sin(\theta) e^{-j\beta r} \tag{D.2}$$

where $\beta = 2\pi/\lambda$ $\tag{D.3}$

These fields may be expressed in terms of the current i in a small conductor of length l by using the relationships:

$$il = \frac{\mathrm{d}p}{\mathrm{d}t}$$

$$= j2\pi fp \tag{D.4}$$

to replace p in the above equations.

At large βr, the only significant fields are:

$$E_\theta = \frac{-p\beta^2}{4\pi\varepsilon_0 r}\sin(\theta)e^{-j\beta r} \tag{D.5}$$

and:

$$H_\phi = \frac{-cp\beta^2}{4\pi r}\sin(\theta)e^{-j\beta r} \tag{D.6}$$

Their ratio is the free-space plane-wave impedance:

$$E_\theta/H_\phi = Z_0 = (\mu_0/\varepsilon_0)^{1/2} = 377\Omega \tag{D.7}$$

where we have used the relationship:

$$c = (\mu_0\varepsilon_0)^{-1/2} \tag{D.8}$$

At small βr the ratio of E to H field is large. The dominant field components are:

$$E_r = \frac{2p}{4\pi\varepsilon_0 r^3}\cos(\theta) \tag{D.9}$$

$$E_\theta = \frac{p}{4\pi\varepsilon_0 r^3}\sin(\theta) \tag{D.10}$$

In general, the wave impedance depends on θ and r. The transverse wave impedance E_θ/H_ϕ is given by:

$$Z_w = Z_0 \frac{\left(\dfrac{1}{\beta^3 r^3} + \dfrac{j}{\beta^2 r^2} - \dfrac{1}{\beta r}\right)}{\left(\dfrac{j}{\beta^2 r^2} - \dfrac{1}{\beta r}\right)} \tag{D.11}$$

For small βr:

$$Z_w = Z_0/(j\beta r)$$

$$= (j2\pi f\varepsilon_0 r)^{-1} \tag{D.12}$$

Similarly, the fields from the oscillatory current i flowing in a small loop of area S, where the dipole moment:

$$m = iS \qquad \text{(D.13)}$$

and:

$$i = i_0 e^{j2\pi ft} \qquad \text{(D.14)}$$

are:

$$E_r = 0$$

$$E_0 = 0$$

$$E_\phi = \frac{-Z_0 m \beta^3}{4\pi} \left(\frac{j}{\beta^2 r^2} - \frac{1}{\beta r} \right) \sin(\theta) e^{-j\beta r}$$

$$H_r = \frac{2m\beta^3}{4\pi} \left(\frac{1}{\beta^3 r^3} + \frac{j}{\beta^2 r^2} \right) \cos(\theta) e^{-j\beta r}$$

$$H_0 = \frac{m\beta^3}{4\pi} \left(\frac{1}{\beta^3 r^3} + \frac{j}{\beta^2 r^2} - \frac{1}{\beta r} \right) \sin(\theta) e^{-j\beta r}$$

$$H_\phi = 0 \qquad \text{(D.15)}$$

At large βr the only significant fields are:

$$E_\phi = \frac{Z_0 m \beta^2}{4\pi r} \sin(\theta) e^{-j\beta r} \qquad \text{(D.16)}$$

and:

$$H_0 = \frac{-m\beta^2}{4\pi r} \sin(\theta) e^{-j\beta r} \qquad \text{(D.17)}$$

and:

$$E_\phi / H_0 = -Z_0 \qquad \text{(D.18)}$$

At small βr, the ratio of E to H field is small. The dominant field components are:

$$H_r = \frac{2m}{4\pi r^3}\cos(\theta) \tag{D.19}$$

$$H_\theta = \frac{m}{4\pi r^3}\sin(\theta) \tag{D.20}$$

In general, the transverse wave impedance E_ϕ/H_θ is given by:

$$Z_w = -Z_0\frac{\left(\dfrac{j}{\beta^2 r^2} - \dfrac{1}{\beta r}\right)}{\left(\dfrac{1}{\beta^3 r^3} + \dfrac{j}{\beta^2 r^2} - \dfrac{1}{\beta r}\right)} \tag{D.21}$$

For small βr:

$$Z_w = -jZ_0\beta r$$

$$= -j2\pi f\mu_0 r \tag{D.22}$$

Note that the dipole orientations are shown in figure 3.1.

The sign of the wave impedance in (D.18) and (D.22) is determined by the orientations of the electric and magnetic field vectors with respect to a coordinate system defined by the source. When used in shielding calculations, the coordinate axes are defined by the field under consideration and positive wave impedances are used.

Exercises

In the following, unless otherwise stated, Boltzman's constant $= 1.38 \times 10^{-23}$ J/K, source temperature $= 290$ K.

Chapter 2

2.1. In the type of pulse-echo distance measurement system shown in figure 1.2, it is important that the threshold is set sufficiently high that the incidence of false triggering by background noise is below a given tolerance level. Show that, if the background noise is Gaussian and at a level of 1 mV rms, then, in the absence of a received echo, the threshold should be set to 2.88 mV for a false triggering rate of 1 for every 500 transmitted pulses.

2.2. If $\mathcal{F}[f(t)] = F(f)$, show that $\mathcal{F}[df(t)/dt] = j2\pi f F(f)$, and that

$$\mathcal{F}\left[\int_{-\infty}^{t} f(\tau)d\tau\right] = F(f)/(j2\pi f).$$

2.3. Show that $\mathcal{F}[f(t)e^{j2\pi f_0 t}] = F(f - f_0)$, and that $\mathcal{F}[f(t - t_0)] = F(f)e^{-j2\pi f t_0}$.

Chapter 3

3.1. In figure 3.6, the input circuit encloses an area of 2 cm^2 and is close to a power-line transformer. The transformer field magnetic induction vector is perpendicular to the plane of the input circuit, varies sinusoidally with a frequency of 50 Hz and has an average amplitude over the area of the circuit of 1 mWb/m^2.
(a) What is the amplitude of the noise from this source at the input of the amplifier?
(b) If the signal from source s is 10 mV rms, what is the available power signal to noise ratio at the output of the amplifier?
(c) How could you improve the signal to noise ratio?

3.2 A signal source with an internal resistance of 10 kΩ is coupled to the input of an amplifier with a voltage gain of 100 and a high input resistance. One side of the source is connected to local ground. The ungrounded PCB track connecting source to amplifier runs close to a power supply track carrying DC and alternating current of power-line frequency (50 Hz) and harmonics.
(a) If the alternating potential difference between the power supply track and local ground has significant components at the following amplitudes and frequencies:

        10 V rms at 50 Hz
         5 V rms at 100 Hz
       2.5 V rms at 150 Hz

and the capacitance between this track and the amplifier input track is 2 pF, calculate and sketch the normalised noise power spectrum at the amplifier output.

(b) If the signal source is 200 μV rms at a frequency of 200 Hz, calculate the available power signal to noise ratio at the output of the amplifier.

(c) Suggest ways of improving this.

3.3. In figure 3.17, the connection betwen B and C has a resistance of 5 mΩ and negligible inductance. The supply current flowing through A, B, C, D has a wideband noise current of $10^{-12} A^2/Hz$ superimposed. The bandwidth of the amplifier is DC – 10 kHz with a sharp high frequency cut-off, and e_s is a sinusoidal signal of 1 kHz at 5 μV rms.

(a) Calculate the signal-to-noise ratio (in dB) at the output of the amplifier.

(b) How would you improve this signal-to-noise ratio?

3.4. Describe five types of electronic noise generated by mechanical movement, together with methods of reduction.

Chapter 4

4.1. An inductance L is connected in parallel with a resistance R. Show that the mean square noise current in L is given by:

$$i_{NL}^2 = \frac{kT}{L}$$

4.2. Calculate the noise bandwidth of a filter having a Gaussian frequency response such that:

$$\frac{|V_{out}|}{|V_{in}|} = A \exp\left[-(f-f_0)^2/(2\sigma^2)\right]$$

where f_0 and σ are constants.

4.3. A capacitor of 0.015 μF and two resistors of 10 kΩ are connected in series such that each component forms one side of a triangle. Calculate the root spectral density in $V\,Hz^{-1/2}$, at a frequency of 1 kHz, of the noise appearing across one of the resistors.

Chapter 5

5.1. An amplifier with a noise bandwidth of 400 Hz centred at 1.0 kHz is to be used with a 0.3 μV rms sine-wave source of frequency 1.0 kHz and resistance 100 Ω. The equivalent input noise generators of the amplifier are $\mathcal{E}_n = 10\,nV\,Hz^{-1/2}$ and $I_n = 10\,pA\,Hz^{-1/2}$.

(a) Calculate the turns ratio of the ideal (lossless) transformer to be used between the source and the amplifier in order to optimise the noise performance.

(b) What is the noise figure (in dB) of the amplifier when used with this transformer?

(c) What is the noise figure (in dB) of the amplifier used without the transformer?

(d) With the transformer in place, calculate the improvement (in dB) of the signal-to-noise ratio at the output of the amplifier when the source is cooled to 77 K using liquid nitrogen. Assume the source resistance and signal output remain unchanged.

5.2. A resistor of value $1.0\,k\Omega$ and an inductor of value $10\,mH$ are connected in series across a $1.0\,mA$ direct current source. A second resistor of value $1.0\,k\Omega$ is connected in parallel with the inductor. The excess noise figure of the resistors is $2\,\mu V\,V_{DC}^{-1}$ (frequency decade)$^{-1/2}$ and the inductor is assumed ideal. Calculate the spectral density (in V^2/Hz) of the noise voltage across the current generator, at a frequency of $15\,kHz$.

5.3. The noise performance of an FET is specified as follows:

Equivalent input noise voltage $= 0.05\,\mu V\,Hz^{-1/2}$
Equivalent input noise current $= 0.15\,pA\,Hz^{-1/2}$

(a) What is the noise figure for this FET when it is used as an amplifier with a source resistance of $100\,k\Omega$?

(b) What is the minimum noise figure and at what source resistance does this occur?

5.4. A sine-wave signal source of $100\,\mu V$ rms at a frequency of $1\,kHz$ with an internal resistance $R_1 = 100\,k\Omega$ is connected to the input of an amplifier with an AC coupled input, an input impedance of $>> 100\,k\Omega$, and a constant voltage gain of 1000 over a frequency range of $1\,Hz$ to $10\,kHz$ with a rapid decrease in gain outside these frequency limits.

The source resistance has an excess noise figure of $0.07\,\mu V$rms V_{DC}^{-1} (frequency decade)$^{-1/2}$. The equivalent input noise generators for the amplifier are uncorrelated and have constant-amplitude spectral densities of $30\,nV\,Hz^{-1/2}$ for the series voltage generator and $0.25\,pA\,Hz^{-1/2}$ for the parallel current generator.

The source (including its internal resistance) has a steady bias current of $200\,\mu A$ flowing through it from an external current generator, and this bias current has superimposed a noise current of constant-amplitude spectral density of $0.4\,pA\,Hz^{-1/2}$.

(a) Calculate the noise level at the output of the amplifier in mV rms.

(b) Calculate the signal-to-noise ratio at the amplifier output in dB.

(c) Briefly describe the steps you could take to improve the signal-to-noise ratio.

Chapter 6

6.1. Derive an expression for the shot noise in a semiconductor junction at zero applied voltage using the Shockley equation relating applied potential difference to current flow, and noting that the two opposing current flows in the equation give rise to independent shot noise currents. Show that this is equal to the noise current of the small-signal AC resistance of the diode at zero applied voltage.

6.2. Draw the low frequency, hybrid-π, common emitter equivalent circuit for a bipolar transistor, incorporating its intrinsic noise sources. Briefly describe and justify this noise model.

6.3. Part of the manufacturer's specification sheet for a particular operational amplifier states:

Parameters	Symbol	Test conditions	Typical values	Units
Input noise voltage density	$\mathcal{E}_n$	$f =$ 1 Hz 10 Hz 100 Hz 1 kHz 10 kHz	100 33.2 10.0 10.0 10.0	$\left.\vphantom{\begin{matrix}1\\2\\3\\4\\5\end{matrix}}\right\}$ nV$\sqrt{}$Hz
Input noise current density	$\mathcal{I}_n$	$f =$ 1 Hz 10 Hz 100 Hz 1 kHz 10 kHz	31.6 10.0 3.32 1.00 1.00	$\left.\vphantom{\begin{matrix}1\\2\\3\\4\\5\end{matrix}}\right\}$ pA/$\sqrt{}$Hz

(a) Assuming that the amplifier noise is dominated by the input stage, say (with reasons) whether this stage is bipolar or FET.

(b) Calculate the minimum noise figure and optimum source resistance at 1 kHz.

(c) If the source signal is a 20 μV rms, 5 kHz sine wave, the source resistance is 100 Ω and the amplifier is followed by a bandpass filter with sharp cut-offs at 1 kHz and 10 kHz, what is the signal-to-noise ratio (in dB) at the output of the filter?

(d) If the lower frequency cut-off is reduced to 1 Hz, what is the signal-to-noise ratio now?

Chapter 7

7.1. Describe the standard method for measuring the average (over the passband) noise factor of an amplifier using a calibrated noise generator of uniform spectral density, a calibrated variable attenuator and an uncalibrated power meter.

7.2. Noise measurements are made on an amplifier with an input impedance of 10 MΩ.

A sine-wave input of 100 μV rms gives an output of 1 V rms at 1 kHz. The noise is negligible at this signal level.

The noise at the output of the amplifier is measured via a bandpass filter with centre frequency 1 kHz, bandwidth 100 Hz, unity gain over the passband, and sharp low and high frequency cut-offs. It is 0.5 mV rms with the input short-circuited and 500 mV rms with the input open-circuit.

Calculate (a) the equivalent input noise current and voltage root spectral densities at 1 kHz, (b) the optimum source resistance and minimum noise figure and (c) the noise figure for source resistances of 1 kΩ, 10 kΩ and 100 kΩ.

7.3. A noise generator with an open-circuit output of 200 nV Hz$^{-1/2}$ and an output impedance of 1 kΩ is connected in series with a 99 kΩ resistor (R_1) and the input of an amplifier. The output of the amplifier is passed to a noise meter with a noise bandwidth of 100 Hz. In the meter, the filtered noise voltage is measured using a squarer followed by a low-pass filter consisting of a series resistor of 100 kΩ and a capacitor to ground followed by a DC meter with high input impedance. The DC meter incorporates a non-linear scale indicating noise voltage referred to the noise meter input. With the noise generator switched off, the noise measured at the amplifier output is 1.08 mV rms. When the noise generator is switched on, the measurement is 2.27 mV rms. When the resistor R_1 is shorted out, the corresponding measurements are 0.108 mV and 2.00 mV respectively.

(a) What is the noise figure of the amplifier with the 100 kΩ source resistance?

(b) What are the amplifier equivalent input noise voltage and current?

(c) What are the optimum source resistance and minimum noise figure?

(d) Calculate the value of the capacitor required in the post-squarer low-pass filter to give an rms error in measured noise voltage of 1 per cent (assume rectangular passband noise filter).

(e) If the noise bandwidth is changed to 10 Hz, what is the measurement error now?

7.4. The excess noise of a resistor is expressed in $\mu V/V_{DC}/\sqrt{\text{frequency}}$ decade. Describe this form of noise, explain the method of expressing it and a method of measuring it.

Chapter 8

8.1. Use SPICE to model the circuit in figure 7.22. Simulate the measurement of the equivalent input noise voltage and current for a number of transistors at various frequencies and operating conditions. How do these compare with the results of chapter 6?

Chapter 9

9.1. (a) Calculate the distance from a small source of electric and magnetic fields to the boundary between the near and far fields at frequencies of (i) 1 MHz and (ii) 100 MHz.

(b) Shielding of coaxial cables and shielding containers improves above the frequency at which the shield thickness equals the skin-depth. If the shield is 0.1 mm thick solid copper, at what frequency does this occur?

(c) The multiple reflection correction term is usually considered negligible for absorptions greater than 15 dB. (i) To what thickness/skin-depth ratio does this correspond and (ii) what is the value of the correction term at this absorption?

(d) The attenuation of a 1 cm diameter circular aperture is to be increased by 64 dB using the waveguide effect. (i) What length of waveguide is required and (ii) over what frequency range could it be used?

9.2. The input circuit of an amplifier forms a loop with an area of 10 cm². The loop is 1 m from, and coplanar with, a power-supply lead carrying a current of 10 A rms at a frequency of 60 Hz. The circuit is to be shielded from the magnetic field of the lead such that the emf induced in the circuit is 20 nV rms. Making the approximations that the shield can be modelled as a sphere of radius 5 cm, and that the magnetic field is uniform over the circuit and the cross-section of the sphere, calculate the thickness of steel ($\mu_r = 10^3$, $\sigma = 0.6 \times 10^7 \, \text{S m}^{-1}$) shield required.

Bibliography and Review Articles

Random signals and signal processing

Bendat, J. P. and Piersol, A. G. (1971). *Random Data: Analysis and Measurement Procedures*, Wiley, New York.

Bracewell, R. N. (1986). *The Fourier Transform and its Application*, 2nd edn, McGraw-Hill, New York.

Davenport, W. B. and Root, W. L. (1958). *Random signals and Noise*, McGraw-Hill, New York.

Lathi, B. P. (1968). *An Introduction to Random Signals and Communication Theory*, International Textbook Co, Scranton, Pennsylvania.

Lynn, P. A. (1992). *Digital Signals, Processors and Noise*, Macmillan, Basingstoke.

Papoulis, A. (1984). *Probability, Random Variables, and Stochastic Processes*, 2nd edn, McGraw-Hill, New York.

Papoulis, A. (1984). *Signal Analysis*, McGraw-Hill, New York.

Intrinsic noise

Ambrozy, A. (1983). *Electronic Noise*, McGraw-Hill, New York.

Baxandall, P. J. (1968). 'Noise in transistor circuits, Part 1', *Wireless World*, **74**, 388–392.

Baxandall, P. J. (1968). 'Noise in transistor circuits, Part 2', *Wireless World*, **74**, 454–459.

Bell, D. A. (1960). *Electrical Noise*, van Nostrand, London.

Bennet, W. R. (1960). *Electrical Noise*, McGraw-Hill, New York.

Buckingham, M. J. (1983). *Noise in Electronic Devices and Systems*, Ellis Horwood, Chichester.

Connor, F. R. (1982). *Noise*, 2nd edn, Edward Arnold, London.

Gupta, M.S. (1988). *Selected Papers on Noise in Circuits and Systems*, IEEE Press, New York.

Hewlett Packard (1983). 'Fundamentals of RF and microwave noise figure measurements', *Application Note 57-1*.

Hewlett Packard (1988). 'Noise figure measurement accuracy', *Application Note 57-2*.

Horowitz, P. and Hill, W. (1989). *Art of Electronics*, 2nd edn, Cambridge University Press (chapter 7).

Letzter, S. and Webster, N. (1970). 'Noise in amplifiers', *IEEE Spectrum*, **7(8)**, 67–75.

Motchenbacher, C. D. and Fitchen, F. C. (1973). *Low-Noise Electronic Design*, Wiley, New York.

Netzer, Y. (1981). 'The design of low-noise amplifiers', *Proceedings of the IEEE*, **69(6)**, 728–742.

Robinson, F. N. (1975). *Noise and Fluctuations in Electronic Devices and Circuits*, Oxford University Press.

van der Ziel, A. (1954). *Noise*, Prentice-Hall, Englewood Cliffs, New Jersey.

van der Ziel, A. (1970). *Noise: Sources, Characterization, Measurement*, Prentice-Hall, Englewood Cliffs, New Jersey.

van der Ziel, A. (1976). *Noise in Measurements*, Wiley, New York.

van der Ziel, A. (1986). *Noise in Solid-State Devices and Circuits*, Wiley, New York.

Electromagnetic compatibility

Baker, D., Koehler, D. C., Fleckenstein, W. O., Roden, C. E. and Sabia, R. (1970). *Physical Design of Electronic Systems. Vol. 1 Design Technology*, Prentice-Hall, Englewood Cliffs, New Jersey (chapter 10).

Benda, S. (1991). *Interference-free Electronics*, Studentlitteratur, Lund, Sweden.

Bunk, D. S. and Donovan, T. J. (1967). 'Electromagnetic shielding', *Machine Design*, **39**, 102–117.

Chatterton, P. A. and Houlden, M. A. (1992). *EMC: Electromagnetic Theory to Practical Design*, Wiley, Chichester.

Ficchi, R. F. (1964). *Electrical Interference*, Hayden, New York.

Ficchi, R. F. (1971). *Practical Designs for Electromagnetic Compatibility*, Hayden, New York.

IEEE (1968). 'Special issue on RF shielding', *IEEE Transactions on Electromagnetic Compatibility*, **EMC-10(1)**.

IEEE (1988). Special issue on electromagnetic shielding, *IEEE Transactions on Electromagnetic Compatibility*, **30(3)**.

Keiser, B. (1987). *Principles of Electromagnetic Compatibility*, 3rd edn, Artech House, Norwood.

Morrison, R. (1986). *Grounding and Shielding Techniques in Instrumentation*, Wiley, New York.

Morrison, R. (1992). *Noise and Other Interfering Signals*, Wiley, New York.

Nalle, D. (1965). 'Elimination of noise in low-level circuits', *Instrument Society of America Journal*, **12**, 59–68.

Ott, H. W. (1988). *Noise Reduction Techniques in Electronic Systems*, Wiley, New York.

Paul, C. R. (1992). *Introduction to Electromagnetic Compatibility*, Wiley, New York.

Rotkiewicz, W. (1982). *Electromagnetic Compatability in Radio Engineering*, Elsevier, Amsterdam.

Severinsen, J. (1975). 'Designer's Guide to EMI Shielding. Part 1', *EDN*, **20**, 47–51.

Severinsen, J. (1975). 'Designer's Guide to EMI Shielding. Part 2', *EDN*, **20**, 53–58.

Smith, A. A. (1977). *Coupling of External Electromagnetic Fields to Transmission Lines*, Wiley, New York.

Standler, R. B. (1989). *Protection of Electronic Circuits from Overvoltages*, Wiley, New York.

Vance, E. F. (1978). *Coupling to Shielded Cables*, Wiley, New York.

Vance, E. F. (1980). 'Electromagnetic-interference control', *IEEE Transactions on Electromagnetic Compatibility*, **EMC-22(4)**, 319–328.

Answers to Exercises

2.1. From (2.17), $\int\limits_{thr}^{\infty} p(x)dx = 1/500$. From tables of area under normal (Gaussian) density function, thr $= 2.88\sigma$.

2.2 and **2.3.** Use equations (2.58) and (2.59).

3.1 (a) 62.83 μV (note that this is amplitude, not rms).
   (b) 0.507×10^5 (47.05 dB).
   (c) (i) Move transformer.
   (ii) Reduce area of input circuit.
   (iii) Reorientate the transformer or input circuit such that the magnetic induction vector is parallel to the plane of the circuit.
   (iv) Change the transformer to a low magnetic leakage type.
   (v) Use magnetic shielding.
   (vi) Filter the amplifier output. For example, use a 50 Hz notch filter if the signal does not include components at or near 50 Hz.

3.2. (a) Line spectrum with components:
   3.95×10^{-5} V^2 at 50 Hz
   3.95×10^{-5} V^2 at 100 Hz
   2.22×10^{-5} V^2 at 150 Hz
   (b) 3.95 (5.97 dB).
   (c) (i) Increase separation of tracks.
   (ii) Put grounded track between the amplifier input track and the power supply track.
   (iii) Replace amplifier input track by screened lead.
   (iv) Remove the ground connection and use a twisted pair to connect the source to a differential input amplifier.
   (v) Use high-pass filter to reject frequencies below 200 Hz.

3.3. (a) 20 dB.
   (b) (i) Use narrowbandpass filter at 1 kHz.
   (ii) Alter the layout so that supply current does not flow through a connection in series with e_s.

4.1. See derivation of (4.14).

4.2. $\sigma\sqrt{\pi}$.

4.3. 9.88×10^{-9} V Hz$^{-1/2}$.

5.1. (a) $1:10^{1/2}$. (b) 11.30 dB. (c) 18.07 dB. (d) 0.24 dB.

5.2. 1.39×10^{-16} V^2/Hz.

5.3. (a) 2.70 (4.3 dB). (b) 1.94 (2.9 dB) at 0.333 MΩ.

5.4. (a) 7.42 mV rms. (b) 22.6 dB.
   (c) (i) Narrowbandpass filter at 1 kHz.
   (ii) Low noise amplifier.
   (iii) Low noise bias current supply.
   (iv) Source resistor with low excess noise figure.

271

(v) Noise match using transformer (difficult at relatively high source impedance and low frequency).

6.3. (a) Bipolar, increase of $\mathcal{I}_n$ at low frequency. (b) 2.25 (3.5 dB), 10 kΩ. (c) 26.4 dB. (d) 25.6 dB.

7.2. (a) 0.5 pA Hz$^{-1/2}$, 5 nV Hz$^{-1/2}$. (b) 10 kΩ, 1.31 (1.18 dB). (c) 2.58 (4.11 dB), 1.31 (1.17 dB), 2.58 (4.11 dB). Note the possibility of calculating the input noise current with an infinite source resistance provided the amplifier voltage gain and input resistance are known.

7.3. (a) 7.31 (8.64 dB). (b) 10 nV Hz$^{-1/2}$ and 1 pA Hz$^{-1/2}$. (c) 10 kΩ and 2.25 (3.52 dB). (d) 125 μF (this illustrates a difficulty of CR averaging with narrow measurement bandwidths – a capacitor of this size may well have a leakage resistance comparable or greater than 100 kΩ). (e) 10$^{1/2}$ per cent.

9.1. (a) (i) 47.7 m, (ii) 0.477 m. (b) 435 kHz. (c) (i) 1.73, (ii) 0.258 dB. (d) (i) 2 cm, (ii) DC to 5.8 GHz ($f_c/3$).

9.2. 2.8 mm.

Index